THE INEQUALITIES

IN THE

MOTION OF THE MOON

DUE TO THE

DIRECT ACTION OF THE PLANETS

THE INEQUALITIES
IN THE
MOTION OF THE MOON
DUE TO THE
DIRECT ACTION OF THE PLANETS

AN ESSAY WHICH OBTAINED THE ADAMS PRIZE
IN THE UNIVERSITY OF CAMBRIDGE FOR THE YEAR 1907

BY

ERNEST W. BROWN Sc.D. F.R.S.

PROFESSOR OF MATHEMATICS IN YALE UNIVERSITY CONNECTICUT
SOMETIME FELLOW OF CHRIST'S COLLEGE CAMBRIDGE AND PROFESSOR OF MATHEMATICS IN HAVERFORD COLLEGE PENNSYLVANIA

CAMBRIDGE
at the University Press
1908

CAMBRIDGE
UNIVERSITY PRESS

University Printing House, Cambridge CB2 8BS, United Kingdom

Published in the United States of America by Cambridge University Press, New York

Cambridge University Press is part of the University of Cambridge.

It furthers the University's mission by disseminating knowledge in the pursuit of education, learning and research at the highest international levels of excellence.

www.cambridge.org
Information on this title: www.cambridge.org/9781107664739

First published 1908
First paperback edition 2014

A catalogue record for this publication is available from the British Library

ISBN 978-1-107-66473-9 Paperback

TO

GEORGE HOWARD DARWIN

AT WHOSE SUGGESTION THE STUDY OF THE MOON'S MOTIONS
WAS UNDERTAKEN BY THE AUTHOR
AND WHOSE ADVICE AND SYMPATHY HAVE BEEN FREELY GIVEN
DURING THE PAST TWENTY YEARS
THIS ESSAY IS GRATEFULLY DEDICATED

CONTENTS.

INTRODUCTION.

THIS Essay aims at a complete calculation of the effects produced by the action of a planet on the motion of the moon under the following limitations and conditions:

(1) The problem of the motion of the moon under the action of the sun (supposed to move round the centre of mass of the earth and moon in a fixed elliptic orbit) and the earth, is considered to have been completely solved.

(2) All the bodies are supposed to attract in the same manner as particles of masses equal to their actual masses and situated at the centres of mass.

(3) All the planets are supposed to move in fixed elliptic orbits, *i.e.*, the effect of the action of a planet transmitted either through the earth or through another planet is neglected.

(4) Perturbations of the first order with respect to the ratio of the mass of a planet to that of the sun are alone calculated.

(5) The exception to the above limitations occurs in the periods of revolution of the apse and node of the moon's orbit. These periods are not exactly those arising from (1) but they are the observed periods or, what amounts to the same thing owing to the close agreement between the observed and calculated periods, the periods after all known causes have been included. The point is only of importance in terms of very long period.

(6) All coefficients greater than $0''\cdot01$ in longitude, latitude and parallax have been obtained. Many are also given which are less than $0''\cdot01$ whenever they have been accurately calculated. There are, in addition, classes of terms of short period which run in series and which in the aggregate will add up at certain times to much more than $0''\cdot01$: these have been found to be $0''\cdot002$.

(7) The maximum period considered is 3500 years, but as the sieve in Section IV. retained a few terms of longer period, these were also included in the general scheme.

The methods here adopted have been constructed mainly to overcome the difficulties which have in the past prevented an accurate computation of the long period terms. They were, however, found to be equally useful for finding the terms which have periods of a year or less. These difficulties include the development of the parts of the disturbing function which depend on the coordinates of the earth and planet; the accurate calculation of the derivatives of the moon's coordinates with respect to n, the moon's mean motion; the uncertainty arising from the possible omission of terms of long period; and the frequent appearance of small coefficients as the difference of two large numbers.

In Section I., the equations of variations for the lunar elements have been recomputed with the use of a semi-canonical system of elements. The equations for the ordinary system of elements were first given by G. W. Hill and independently, though at a later date, by S. Newcomb; they were recalculated in Hill's form by R. Radau. In the present system, errors arising from the slow convergence of Delaunay's series have been avoided; in fact his literal expressions have only been used in small terms where derivatives with respect to n were required but where the maximum possible error could make no difference in the final results.

In Section II. Hill's method of dividing the disturbing function into a sum of products in which the first factor of each product is independent of the lunar coordinates and the other of the planet's coordinates, is exhibited as part of a general theorem.

By referring these coordinates to the *true* place of the sun's radius vector, I have obtained the first factors directly from the expansion of the inverse first power of the distance between the planet and the earth ($1/\Delta$) and of its derivatives with respect to certain of the elements of the earth and planet. Only one expansion is therefore required for all these factors, namely, that of $1/\Delta$, and this has been given by Leverrier in a literal form in powers of the eccentricities and mutual inclination. The expansion also contains the coefficients in the expansion of $1/\Delta_0$ (the value of a'/Δ when the eccentricities and inclination are zero) and its derivatives with respect to the planet's mean distance: the formulae for finding the coefficients and their derivatives are here put into forms which admit of rapid and simple computation.

The factors containing the moon's coordinates, together with their derivatives with respect to all the lunar elements except n, are found from the results of my lunar theory; a special method which I gave five years ago and which does not require the use of literal series in powers of n'/n has been used to find the derivatives with respect to n. These methods for finding the planetary and lunar factors are set forth in Section III.

The only one of the difficulties previously mentioned which has not been considered up to this point is the danger of omitting long-period terms with sensible coefficients. In Section IV. formulae are constructed which permit one to find rapidly an upper limit to the magnitude of any coefficient. By means of them, all terms having coefficients greater than 0″·01 and periods less than 3500 years have been sifted out; there are about 100 of such terms, excluding the terms of short period for which no sieve was required.

Sections V., VI. consist of numerical results. It may be noted that the values of A_p, B_p, ... are also required for finding the perturbations of the earth by the planets and of the planets by the earth, while those of M_i and of their derivatives (as well as the equations of variations contained in Section I.) are available for the computation of lunar perturbations other than those due to the direct action of the planets.

No new inequalities sufficiently great to account for the observed discrepancy between theory and observation appear from the direct action of the planets, as shown in the tables of Section VI. Radau's well-known list of terms in longitude has required considerable extension as far as the short period terms are concerned, and a few new long period inequalities with small coefficients have been computed. The more extensive developments of this essay have shown that some of his coefficients require alteration, but there is a general agreement for all those portions which he has taken into account.

Only a few slight verbal changes and corrections of errors in copying have been made to the first five sections, with two exceptions mentioned below, since the award of the examiners. But I have gone over all the computations for finding the short period terms and the larger long period terms during the year that has elapsed and have made the following corrections to the results in Section VI.:

Argument	Former coefficient	Corrected coefficient
$l+3T-10V+33°$	− 0″·35	+ 0″·35
$-l-16T+18V-151°$	−15″·22	−14″·55
$l+29T-26V+112°$	+ 0″·117	+ 0″·108
$2D-l+21T-20V-87°$	+ 0″·111	+ 0″·126
$2l-2D+6M-5T+211°-l$	+ 0″·040	− 0″·038

together with their accompanying short period secondaries.

The signs of those coefficients containing h on p. 86 have been changed.

The annual mean motion of the perigee has been altered from 2″·66* to 2″·69.

* I gave this value in a paper referred to on page 3 below.

The values of M_2 and of its derivatives with respect to n, e, k under the argument 0 on page 61 have required a factor 2. This error necessitated slight changes in some of the coefficients whose primaries were independent of the lunar angles; the largest correction was one of $-0''\cdot 019$ in the coefficient of $\sin(T-V)$.

A wrong sign in the computation of the equations of variations (see Errata at the end of the volume) gave rise to a few almost insensible changes in certain coefficients.

The additions are:

Argument	Coefficient
$4M - 2T + 63°$	$-0''\cdot 012$
$2l - 2\text{D} + 4(M - T) - l$	$+0''\cdot 017$
$2l - 2\text{D} + 8M - 6T + 63°$	$-0''\cdot 019$
$2l - 2\text{D} + 8M - 6T + 63° - l$	$-0''\cdot 031$

No other change or additional coefficient has been greater than $0''\cdot 010$.

The Addendum containing the results obtained by adding together terms of the same argument in Section VI. is also new.

During the summer of last year Professor Newcomb's new work* on this subject was published. His methods differ so completely from those given here that no comparison is made easily except in a few of the final results where the indirect action is insensible or is separated from the direct action. For the large inequality due to Venus he obtains a coefficient of $14''\cdot 83$ while mine is $14''\cdot 55$; a portion of the difference is probably due to certain terms of the second order relative to the ratios of the masses of Venus and of the earth to the sun which Professor Newcomb has included. On the other hand, he states that the possible errors arising in his method may be of the order of this difference, while such errors are excluded from my result. His results and mine for the annual mean motions of the perigee and node agree within $0''\cdot 01$, which is the limit of accuracy to which I have obtained these quantities.

* "Investigation of Inequalities in the Motion of the Moon produced by the Action of the Planets." Carnegie Institution, *Publication* 72, Washington, D.C., June, 1907.

E. W. B.

NEW HAVEN, CONN., U.S.A.
1908 *March* 27.

GENERAL NOTATION.

THE axes of x, y are taken in the ecliptic of 1850·0 and the centre of mass of the earth and the moon is supposed to describe a fixed ellipse around the sun in this plane. As it is more convenient to use the motion of the centre of the earth than of this centre of mass, a slight well-known change, noted below, is necessary in the disturbing function.

The axis of x is parallel to the line joining the earth and the sun.

In the scheme of notation which follows, two sets of constants are given for the mean distance, eccentricity and sine of half the inclination of the moon's orbit. The first set is that which I have used in the expressions for the rectangular coordinates of the moon; the second set is that of Delaunay in the final form which he gives to the expressions for the longitude, latitude and parallax. The longitudes of a *planet,* of its perigee and of its node are as usual reckoned along the ecliptic to its node and then along the orbit.

	Moon	Earth	Planet
True long.	V	V'	V''
Mean long.	$w_1 = l + g + h$	T	P
Mean anom.	$w_1 - w_2 = l$	$l' = T - \varpi'$	$l'' = P - \varpi''$
Mean long. of node	$w_3 = h$	0	h''
Mean motion	n	n'	$\frac{dP}{dt}$
Mean distance	a, a	a'	a''
Eccentricity	e, e	e'	e''
Sine half inclin.	k, γ	0	γ''
Coors., origin earth	x, y, z, r		ξ, η, ζ, Δ
„ „ sun		x', y', z', r'	x'', y'', z'', r''

$$n = \text{mean motion of the moon} = \frac{dw_1}{dt},$$

$$b_2 = \text{„ „ its perigee} = \frac{dw_2}{dt},$$

$$b_3 = \text{„ „ „ node} = \frac{dw_3}{dt},$$

c_1, c_2, c_3 are the canonical constants complementary to w_1, w_2, w_3 *after* the problem of the moon's motion as disturbed by the sun, supposed to move in a fixed elliptic orbit, has been solved.

R = the disturbing function of this problem, arising from the direct attraction of a planet.

The symbols for the mean longitudes of the planets are: Mercury, Q; Venus*, V; Mars, M; Jupiter, J; Saturn, S.

* No confusion will be caused by the use of the same symbol for the true longitude of the moon and the mean longitude of Venus. The notation of Radau has been adopted with a few changes.

SECTION I.

THE EQUATIONS OF VARIATIONS.

LET w_1, w_2, w_3 represent the mean longitudes of the moon, of its perigee and of its node, and suppose the problem of the moon as disturbed by the sun, has been solved. Then it is well known that if a disturbing function R be added to the force function of the moon's motion, the change in the latter due to R can be obtained by solving the equations

$$(1)\quad \frac{dc_i}{dt}=\frac{\partial R}{\partial w_i},\quad \frac{dw_i}{dt}=-\frac{\partial R}{\partial c_i}+b_i \qquad (i=1, 2, 3),$$

where b_1, b_2, b_3 are the mean motions of the moon, of its perigee and of its node, and c_1, c_2, c_3 are the canonical constants corresponding to w_1, w_2, w_3; the c_i are functions of the arbitrary constants n, e, k of the moon's motion, and they also contain the constants n', e', depending on the sun's motion. The substitution of the new values of c_i, w_i, thus found, in the expressions for the coordinates will give the disturbed position of the moon at any time.

The constant part of R only gives constant additions to the b_i, i.e., to the mean motions*: this part will be neglected, since it has no effect on the new terms to be found. Hence $b_1 = n$.

Change to the semi-canonical system n, c_2, c_3, retaining the w_i unchanged. Putting

$$\frac{dc_1}{dn}=-a^2\beta,$$

and remembering that

$$(1\text{a})\quad \frac{dc_1}{dc_2}=-\frac{db_2}{dn},\quad \frac{dc_1}{dc_3}=-\frac{db_3}{dn},\quad \frac{db_2}{dc_3}=\frac{db_3}{dc_2},$$

* I have found these changes in an earlier paper: *Trans. Amer. Math. Soc.*, Vol. v. pp. 279—284. A fresh computation just made gives 2″·69, −1″·42, for the mean motions of the perigee and node respectively. The former is 0″·03 more than the value given in the paper.

we obtain equations (1) in the semi-canonical form

$$(2)\quad \begin{cases} \dfrac{dn}{dt} = \dfrac{1}{a^2\beta}\left(-\dfrac{\partial R}{\partial w_1} - \dfrac{db_2}{dn}\cdot\dfrac{dc_2}{dt} - \dfrac{db_3}{dn}\cdot\dfrac{dc_3}{dt}\right), & \dfrac{dw_1}{dt} = \dfrac{1}{a^2\beta}\dfrac{\partial R}{\partial n} + b_1, \\ \dfrac{dc_2}{dt} = \dfrac{\partial R}{\partial w_2}, & \dfrac{dw_2}{dt} = -\dfrac{\partial R}{\partial c_2} + b_2 + \left(\dfrac{dw_1}{dt} - b_1\right)\dfrac{db_2}{dn}, \\ \dfrac{dc_3}{dt} = \dfrac{\partial R}{\partial w_3}, & \dfrac{dw_3}{dt} = -\dfrac{\partial R}{\partial c_3} + b_3 + \left(\dfrac{dw_1}{dt} - b_1\right)\dfrac{db_3}{dn}; \end{cases}$$

in these equations b_2, b_3, c_1 are supposed to be expressed in terms of n, c_2, c_3 and R in terms of n, c_2, c_3, w_1, w_2, w_3.

Consider any periodic term of R:

$$R = n'^2 \mathrm{a}^2 A \cos(qt + q') = n'^2 \mathrm{a}^2 A \cos(i_1 w_1 + i_2 w_2 + i_3 w_3 + q''t + q'''),$$

where a is the linear constant of Hill's variational orbit and of my lunar theory and A is a numerical coefficient (that is, its dimensions with respect to time, space and mass are zero); $q''t + q'''$ is a combination of the solar and planetary arguments. Then since $qt + q'$ is independent of the c_i and A of the w_i, the first three of equations (2) become

$$\frac{dn}{dt} = \frac{n'^2}{\beta}\cdot\frac{\mathrm{a}^2}{a^2} A \frac{dq}{dn}\sin(qt + q'),$$

$$\frac{dc_2}{dt} = -i_2 n'^2 \mathrm{a}^2 A \sin(qt + q'),$$

$$\frac{dc_3}{dt} = -i_3 n'^2 \mathrm{a}^2 A \sin(qt + q').$$

It will now be supposed that R contains a small factor whose square may be neglected. The coefficients in the right-hand members of the last set will then be constants and we can integrate. Put $m = \dfrac{n'}{n}$, and let δn, δc_2, δc_3 denote the increments of n, c_2, c_3, due to R. Then

$$(3)\quad \begin{cases} \dfrac{\delta n}{n} = -\dfrac{m}{\beta}\cdot\dfrac{\mathrm{a}^2}{a^2}\cdot\dfrac{dq}{dn}\dfrac{n'}{q} A\cos(qt + q'), \\ \dfrac{\delta c_2}{na^2} = i_2 m \,.\, \dfrac{\mathrm{a}^2}{a^2}\dfrac{n'}{q} A\cos(qt + q'), \\ \dfrac{\delta c_3}{na^2} = i_3 m \,.\, \dfrac{\mathrm{a}^2}{a^2}\dfrac{n'}{q} A\cos(qt + q'). \end{cases}$$

Again, if we put

$$A_1 = \frac{n}{\mathrm{a}^2}\frac{d}{dn}(\mathrm{a}^2 A),$$

$$A_2 = -a^2 n \frac{dA}{dc_2},$$

$$A_3 = -a^2 n \frac{dA}{dc_3},$$

the other three of equations (2) become

$$
(3\mathrm{a})\quad \begin{cases} \dfrac{dw_1}{dt} = \dfrac{n'^2}{n\beta} \cdot \dfrac{\mathrm{a}^2}{a^2} A_1 \cos(qt+q') + b_1, \\[2ex] \dfrac{dw_2}{dt} = \dfrac{n'^2}{n} \cdot \dfrac{\mathrm{a}^2}{a^2} \left(A_2 + \dfrac{A_1}{\beta} \cdot \dfrac{db_2}{dn}\right) \cos(qt+q') + b_2, \\[2ex] \dfrac{dw_3}{dt} = \dfrac{n'^2}{n} \cdot \dfrac{\mathrm{a}^2}{a^2} \left(A_3 + \dfrac{A_1}{\beta} \cdot \dfrac{db_3}{dn}\right) \cos(qt+q') + b_3. \end{cases}
$$

Let δb_1, δb_2, δb_3, δw_1, δw_2, δw_3 denote increments due to R. Then

$$
\delta b_1 = \delta n = -\frac{n'}{\beta} \cdot \frac{\mathrm{a}^2}{a^2}\frac{dq}{dn} \cdot \frac{n'}{q} A \cos(qt+q'),
$$

$$
\begin{aligned}
\delta b_2 &= \frac{db_2}{dn}\delta n + \frac{db_2}{dc_2}\delta c_2 + \frac{db_2}{dc_3}\delta c_3 \\
&= n' \frac{\mathrm{a}^2}{a^2} \cdot \frac{n'}{q} A \left(-\frac{1}{\beta} \cdot \frac{db_2}{dn} \cdot \frac{dq}{dn} + a^2 \frac{db_2}{dc_2} \cdot i_2 + a^2 \frac{db_2}{dc_3} \cdot i_3\right) \cos(qt+q'),
\end{aligned}
$$

by equations (1a).

Putting $$q_2 = -na^2 \frac{dq}{dc_2}, \quad q_3 = -na^2 \frac{dq}{dc_3},$$

and using the last of equations (1a), we find

$$
\delta b_2 = -n' \frac{\mathrm{a}^2}{a^2} \cdot \frac{n'}{q} A \left(\frac{1}{\beta} \cdot \frac{db_2}{dn} \cdot \frac{dq}{dn} + q_2\right) \cos(qt+q');
$$

similarly, $$\delta b_3 = -n' \frac{\mathrm{a}^2}{a^2} \cdot \frac{n'}{q} A \left(\frac{1}{\beta} \cdot \frac{db_3}{dn} \cdot \frac{dq}{dn} + q_3\right) \cos(qt+q').$$

The undisturbed values of b_1, b_2, b_3 are of course equal to those of $\dfrac{dw_1}{dt}$, $\dfrac{dw_2}{dt}$, $\dfrac{dw_3}{dt}$, respectively. To obtain δw_1, δw_2, δw_3 (the increments of w_1, w_2, w_3), it is necessary to replace w_i by $w_i + \delta w_i$, and b_i by $b_i + \delta b_i$ in (3a), to substitute for δb_i the values just obtained, and then to integrate. These operations give

$$
(4)\quad \begin{cases} \delta w_1 = \dfrac{1}{\beta}\dfrac{\mathrm{a}^2}{a^2}\left(m\dfrac{n'}{q}A_1 - \dfrac{dq}{dn}\dfrac{n'^2}{q^2}A\right)\sin(qt+q'), \\[2ex] \delta w_2 = \dfrac{\mathrm{a}^2}{a^2}\left\{m\dfrac{n'}{q}\left(A_2 + \dfrac{A_1}{\beta}\dfrac{db_2}{dn}\right) - \dfrac{n'^2}{q^2}A\left(\dfrac{q_2}{n} + \dfrac{1}{\beta}\dfrac{db_2}{dn}\dfrac{dq}{dn}\right)\right\}\sin(qt+q'), \\[2ex] \delta w_3 = \dfrac{\mathrm{a}^2}{a^2}\left\{m\dfrac{n'}{q}\left(A_3 + \dfrac{A_1}{\beta}\dfrac{db_3}{dn}\right) - \dfrac{n'^2}{q^2}A\left(\dfrac{q_3}{n} + \dfrac{1}{\beta}\dfrac{db_3}{dn}\dfrac{dq}{dn}\right)\right\}\sin(qt+q'). \end{cases}
$$

The equations (3), (4) constitute the solution of the problem.

The form in which a periodic term of R arises (see Section II.) is

$$
R = \frac{1}{4}\frac{m''}{m'} n'^2 \mathrm{a}^2 A \cos(qt+q').
$$

It is therefore necessary to multiply the right-hand members of (3), (4) by $m''/4m'$.

Next, let

s' = no. of seconds in the daily mean motion of the sun $= 3548''\cdot 19$,

s = " " " " argument $qt+q'$.

Then
$$\frac{n'}{q}=\frac{s'}{s}.$$

Also, put for brevity

$$(5)\quad \begin{cases} f=\dfrac{1}{4}\dfrac{m''}{m'}\dfrac{\mathrm{a}^2}{a^2}\dfrac{s'^2}{\beta}\,206265, \\[2ex] f'=\dfrac{1}{4}\dfrac{m''}{m'}\dfrac{\mathrm{a}^2}{a^2}\dfrac{ms'}{\beta}\,206265. \end{cases}$$

The coefficients of the right-hand members being thus expressed in seconds of arc, equations (3), (4) become

$$(6)\quad \begin{cases} \dfrac{\delta n}{n}=-f'\dfrac{dq}{dn}\dfrac{A}{s}\cos(qt+q'), \\[2ex] \dfrac{\delta c_2}{na^2}=i_2\beta f'\dfrac{A}{s}\cos(qt+q'), \\[2ex] \dfrac{\delta c_3}{na^2}=i_3\beta f'\dfrac{A}{s}\cos(qt+q'), \\[2ex] \delta w_1=\left(f'\dfrac{A_1}{s}-f\dfrac{A}{s^2}\dfrac{dq}{dn}\right)\sin(qt+q'), \\[2ex] \delta w_2=\left\{\dfrac{f'}{s}\left(\beta A_2+\dfrac{db_2}{dn}A_1\right)-\dfrac{f}{s^2}A\left(\beta\dfrac{q_2}{n}+\dfrac{db_2}{dn}\cdot\dfrac{dq}{dn}\right)\right\}\sin(qt+q'), \\[2ex] \delta w_3=\left\{\dfrac{f'}{s}\left(\beta A_3+\dfrac{db_3}{dn}A_1\right)-\dfrac{f}{s^2}A\left(\beta\dfrac{q_3}{n}+\dfrac{db_3}{dn}\cdot\dfrac{dq}{dn}\right)\right\}\sin(qt+q'); \end{cases}$$

where, to recall certain definitions,

$$\beta=-\frac{1}{a^2}\cdot\frac{dc_1}{dn},\quad q=i_1n+i_2b_2+i_3b_3+q'',$$

$$q_2=-na^2\frac{dq}{dc_2},\quad q_3=-na^2\frac{dq}{dc_3}.$$

Numerical form of the equations of variations.

It remains to be seen how these quantities may be put into numerical form. In f, f' the factor $\dfrac{\mathrm{a}^2}{a^2}$ is immediately obtained from Hill's results* for the variational orbit; m is a well known quantity; $\dfrac{m''}{m'}$ is known as soon as

* *Amer. Jour. Math.*, Vol. I. p. 249.

the particular planet is chosen; thus f, f' remain the same for a given planet and f/f' for all planets. The coefficients A, A_1, A_2, A_3 will be found later on, while q is known as soon as the particular term of R has been chosen.

There remain for calculation

$$(7) \quad \frac{dc_1}{dn}, \; \frac{db_2}{dn}, \; \frac{db_3}{dn}, \; \frac{db_2}{dc_2}, \; \frac{db_2}{dc_3} = \frac{db_3}{dc_2}, \; \frac{db_3}{dc_3},$$

which, depending only on the orbit of the moon as attracted by the sun and earth, are the same for every perturbation of the moon's motion, and therefore apply not only to the present investigation but also to all investigations where a disturbing function R is added to the moon's force function.

Some idea of the degree of accuracy required is desirable. The largest known inequality is that with the argument $l + 16T - 18V$, which has a coefficient of about 15″. For this $i_1 = 1$, $i_2 = -1$, $i_3 = 0$. The principal part is given by

$$-f\frac{A}{s^2}\frac{dq}{dn} = -f\frac{A}{s^2}\left(1 - \frac{db_2}{dn}\right) = -f\frac{A}{s^2}(1 + \cdot 01486).$$

There is no other coefficient which is so great as 2″. Since the degree of accuracy aimed at is 0″·01, it will be sufficient to use four place logarithms and four significant figures for the functions (7) so that the final results will be accurate to at least three significant figures. But certain of the functions are only needed to one or two significant figures, as will appear immediately.

The functions c_1, c_2, c_3 are the same as Delaunay's L, $G - L$, $H - G$ after the final transformations and the changes to his final system of arbitraries, n, e, γ have been made. As my results will be used for the calculation of the moon functions, it will be more convenient to transfer A_2, A_3 to my constants e, k.

Let $\frac{dQ}{dn}$ denote the derivative of a function Q with respect to n when it is expressed in terms of n, c_2, c_3, and $\left(\frac{dQ}{dn}\right)$ when it is expressed in terms of n, e^2, γ^2. Then the following equations serve for the transformation of the derivatives of Q from one set to the other*:

$$(8) \quad \begin{cases} \dfrac{dQ}{dn} = \left(\dfrac{dQ}{dn}\right) - \dfrac{dQ}{dc_2} \cdot \left(\dfrac{dc_2}{dn}\right) - \dfrac{dQ}{dc_3}\left(\dfrac{dc_3}{dn}\right), \\[2ex] \dfrac{dQ}{dc_2} = \left[\dfrac{dQ}{de^2} - \dfrac{dQ}{dc_3}\dfrac{dc_3}{de^2}\right] \div \dfrac{dc_2}{de^2}, \\[2ex] \dfrac{dQ}{dc_3} = \left[\dfrac{dQ}{d\gamma^2} - \dfrac{dQ}{dc_2} \cdot \dfrac{dc_2}{d\gamma^2}\right] \div \dfrac{dc_3}{d\gamma^2}; \end{cases}$$

* The functions considered here involve e, γ only in the even powers.

the same equations will serve for the transformation from the set n, c_2, c_3 to the set n, e², k² if we replace e^2, γ^2 by e², k², respectively.

It is first to be noticed that

$$\begin{aligned} c_2 &= na^2e^2 \left\{-\tfrac{1}{2} + \text{power series in } m,\, e^2,\, \gamma^2,\, e'^2,\, \left(\frac{a}{a'}\right)^2\right\}, \\ c_3 &= na^2\gamma^2 \{-2 + \quad \text{,,} \quad \text{,,} \quad \text{,,} \quad \text{,,} \quad \}, \\ c_1 &= na^2 \{\ 1 + \quad \text{,,} \quad \text{,,} \quad \text{,,} \quad \text{,,} \quad \}, \\ b_2 &= nm^2 \{\ \tfrac{3}{4} + \quad \text{,,} \quad \text{,,} \quad \text{,,} \quad \text{,,} \quad \}, \\ b_3 &= nm^2 \{-\tfrac{3}{4} + \quad \text{,,} \quad \text{,,} \quad \text{,,} \quad \text{,,} \quad \}; \end{aligned}$$

the power series in each case vanishing with m. Hence when for Q in equations (8) is put c_1, b_2 or b_3, the principal part of the first term in each equation will have a portion independent of e^2, γ^2, while the other terms will have one of these quantities as a factor. Since $e = \frac{1}{18}$, $\gamma = \frac{1}{22}$ approximately, two significant figures will be fully sufficient for the values of $\frac{dc_2}{dn}, \frac{dc_3}{dn}$. It is to be noted that the second and third of equations (8) do not depend on derivatives with respect to n, and therefore that we may change to the variables e, γ or to e, k, from c_2, c_3 without reference to the first equation.

From Newcomb's transformation* of Delaunay's values for L, G, H, we have

Terms in	$\frac{c_2}{na^2e^2}$	$\frac{1}{a^2e^2}\left(\frac{dc_2}{dn}\right)$	Terms in	$\frac{c_3}{na^2\gamma^2}$	$\frac{1}{a^2\gamma^2}\left(\frac{dc_3}{dn}\right)$
m^0	− ·50015	+ ·1667	m^0	− 1·99704	+ ·6657
m^2	+ ·01686	− ·0393	m^2	− ·00322	+ ·0075
m^3	+ ·00644	− ·0215	m^3	+ ·00049	− ·0016
m^4	+ ·00170	− ·0074	m^4	+ ·00005	− ·0002
rem.	+ ·00032	− ·0020	rem.	− ·00022	+ ·0010
Sum	− ·47483	+ ·0965	Sum	− 1·99994	+ ·6724

The remainders in the first and third columns are obtained from the values of c_2, c_3 (calculated by processes independent of the nature of convergence along powers of m) given by myself†; those in the second and fourth columns are estimated from them. In any case the results for $\left(\frac{dc_2}{dn}\right)$, $\left(\frac{dc_3}{dn}\right)$ are correct within two per cent. and this is all the accuracy necessary for our purpose.

In a similar manner the values of $\left(\frac{db_2}{dn}\right)$, $\left(\frac{db_3}{dn}\right)$ can be obtained with

* "Action of the Planets on the Moon," *Amer. Eph. Papers*, Vol. v. Pt 3, pp. 201, 202.

† "Theory of the Motion of the Moon," *Trans. R. A. S.*, Vol. LVII. pp. 64, 65.

more than needful accuracy from the series in powers of m given by Delaunay*, Hill†, and Adams‡ with my numerical values§. The derivatives with respect to a^2, γ^2 are obtained immediately from the last named reference. The derivatives with respect to e^2, k^2 can be immediately derived by inserting the values of $\frac{e}{\mathrm{e}}, \frac{\gamma}{\mathrm{k}}$ which I have also given§: the change in the derivatives with respect to n is insensible.

I find the following results to four places in the logarithms, to each of which 10 has been added:

$$\left(\frac{dc_2}{dn}\right) = +[6{\cdot}4637]\, a^2, \quad \frac{dc_2}{d\mathrm{e}^2} = -[9{\cdot}0775]\, na^2, \quad \frac{dc_2}{d\mathrm{k}^2} = -[7{\cdot}5002]\, na^2,$$

$$\left(\frac{dc_3}{dn}\right) = +[7{\cdot}1313]\, a^2, \quad \frac{dc_3}{d\mathrm{e}^2} = +[6{\cdot}7577]\, na^2, \quad \frac{dc_3}{d\mathrm{k}^2} = -[10{\cdot}3039]\, na^2,$$

$$\left(\frac{db_2}{dn}\right) = -[8{\cdot}1709], \quad \frac{db_2}{d\mathrm{e}^2} = -[7{\cdot}3974]\, n, \quad \frac{db_2}{d\mathrm{k}^2} = -[8{\cdot}7004]\, n,$$

$$\left(\frac{db_3}{dn}\right) = +[7{\cdot}5736], \quad \frac{db_3}{d\mathrm{e}^2} = -[7{\cdot}4731]\, n, \quad \frac{db_3}{d\mathrm{k}^2} = +[7{\cdot}8692]\, n.$$

Whence, accurately to four figures,

$$\frac{db_2}{dn} = -[8{\cdot}1720], \quad a^2\frac{db_2}{dc_2} = +[8{\cdot}3175], \quad a^2\frac{db_2}{dc_3} = +[8{\cdot}3960],$$

$$\frac{db_3}{dn} = +[7{\cdot}5733], \quad a^2\frac{db_3}{dc_2} = +[8{\cdot}3960], \quad a^2\frac{db_3}{dc_3} = -[7{\cdot}5698].$$

We have also

$$\frac{\mathrm{a}^2}{a^2} = +[9{\cdot}99921], \quad m = +[8{\cdot}87391], \quad \beta = +[9{\cdot}51801],$$

$$f = [22{\cdot}29358]\frac{m''}{m'}, \quad f' = [17{\cdot}61748]\frac{m''}{m'}, \quad \frac{f'}{f} = [5{\cdot}32390],$$

β being found accurately in Section III. From these quantities the terms depending on $\frac{1}{s^2}$ in (4) can be found.

Further, by substituting A_2, A_3 successively for Q in the second and third of equations (8), and putting e for e, and k for γ, we find

$$A_2 = -a^2n\frac{dA}{dc_2} = [11{\cdot}5819]\frac{dA}{d\mathrm{e}} - [8{\cdot}4242]\frac{dA}{d\mathrm{k}},$$

$$A_3 = -a^2n\frac{dA}{dc_3} = [10{\cdot}7440]\frac{dA}{d\mathrm{k}} - [8{\cdot}7782]\frac{dA}{d\mathrm{e}}.$$

* *Comptes Rendus*, Vol. LXXIV. pp. 19 *et sqq.*

† *Acta Math.*, Vol. VIII. pp. 1—36. *Annals of Math.*, Vol. IX. pp. 31—41.

‡ *Monthly Notices, R. A. S.*, Vol. XXXVIII. pp. 43—49.

§ *Loc. cit.*, Vol. LXIV. p. 532.

Finally, making these various substitutions in (3), (4), the equations of variations become

$$(9)\quad \begin{cases} \dfrac{\delta n}{n} = (-i_1 + \cdot 01486 i_2 - \cdot 003744 i_3) f' \dfrac{A}{s} \cos(qt + q'), \\[2mm] \dfrac{\delta c_2}{na^2} = +[9\cdot 51801]\, i_2 f' \dfrac{A}{s} \cos(qt + q'), \\[2mm] \dfrac{\delta c_3}{na^2} = +[9\cdot 51801]\, i_3 f' \dfrac{A}{s} \cos(qt + q'); \\[2mm] \delta w_1 = \left\{(-i_1 + \cdot 01486 i_2 - \cdot 003744 i_3) f \dfrac{A}{s^2} + f' \dfrac{A_1}{s}\right\} \sin(qt + q'), \\[2mm] \delta w_2 = \left\{(+\cdot 01486 i_1 - \cdot 007066 i_2 - \cdot 008148 i_3) f \dfrac{A}{s^2} \right. \\ \qquad \left. + \left(-[8\cdot 1720]\, A_1 + [11\cdot 0999] \dfrac{dA}{de} - [7\cdot 9422] \dfrac{dA}{dk}\right) \dfrac{f'}{s}\right\} \sin(qt + q'), \\[2mm] \delta w_3 = \left\{(-\cdot 003744 i_1 - \cdot 008148 i_2 + \cdot 001210 i_3) f \dfrac{A}{s^2} \right. \\ \qquad \left. + \left(+[7\cdot 5733]\, A_1 + [10\cdot 2620] \dfrac{dA}{dk} - [8\cdot 2962] \dfrac{dA}{de}\right) \dfrac{f'}{s}\right\} \sin(qt + q'), \end{cases}$$

where
$$A_1 = \frac{n}{a^2} \frac{d}{dn} (A a^2),$$

A being expressed in terms of n, c_2, c_3; the figures enclosed in square brackets being, as elsewhere, logarithms with 10 added.

These equations replace the equations of variations for $\delta l, \delta g, \delta h, \delta a, \delta e, \delta \gamma$ first given by G. W. Hill* and calculated also by Radau†. To reduce the above system to theirs it is necessary to find $\delta c_2, \delta c_3$ in terms of $\delta n, \delta e, \delta \gamma$. But it is simple to compare the first terms of δw_1, on which the principal part of each long-period inequality depends. Radau finds

$$\delta w_1 = \{(-3\cdot 0576 i + \cdot 05601 i' - \cdot 01124 i'') p + \ldots\} P \sin(qt + q'),$$

where $\quad i_1 = i + i' + i'', \quad i_2 = i' + i'', \quad i_3 = i'', \quad pP = \beta f \dfrac{A}{s^2}.$

This reduction gives

$$\delta w_1 = \left\{(-\cdot 9894 i_1 + \cdot 01476 i_2 - \cdot 003705 i_3) f \frac{A}{s^2} + \ldots\right\} \sin(qt + q'),$$

which is about $\frac{1}{100}$ less than my value. But I have been unable to find the coefficient $-3\cdot 0576 i$ in δw_1, which Radau gives: from his data I make this coefficient $-3\cdot 0791$‡ which, reduced to my form, becomes $-\cdot 9964$. The

* *American Eph. Papers*, Vol. III. p. 390.

† *Loc. cit.*, pp. 35, 36.

‡ This apparent error seems to be due to some confusion in the substitution of the numerical values for n before and after the final Delaunay transformation.

difference, ·0036, is about the error to be expected owing to the slow convergence of the series from which Radau obtains his coefficient. It may be added that the value of β, on which my coefficient of i_1 alone depends in this connection, can be determined accurately to at least seven significant figures, and the value for it quoted above has been again verified and tested.

The Variation of the Moon's True Longitude.

The coefficients in longitude are obtained by substituting the values thus found in

$$(10)\quad \delta V = \delta w_1 + \left(\frac{dV}{dw_1} - 1\right)\delta w_1$$
$$+ \frac{dV}{dw_2}\,.\,\delta w_2 + \frac{dV}{dw_3}\,.\,\delta w_3 + \frac{dV}{dn}\,.\,\delta n + \frac{dV}{dc_2}\,.\,\delta c_2 + \frac{dV}{dc_3}\,.\,\delta c_3$$

the first term of which contains the primary inequalities, and the remaining terms the accompanying secondary inequalities.

The largest coefficient in V—that of $\sin l$—produces a maximum coefficient in δV through $\frac{dV}{dc_2}\,.\,\delta c_2$, less than 1″; the coefficient of $\sin l$ in V is 22640″. It has therefore been necessary, in order to include all coefficients greater than 0″·005, to find those terms in V having the factor e^p which have coefficients greater than 100″, or, if $p = 0$, to 400″.

From my results I obtain

$$V = w_1 + [9{\cdot}0405]\sin l + [8{\cdot}3470]\sin(2D - l)$$
$$+ [6{\cdot}9688]\sin(2D + l) + [6{\cdot}2703]\sin(4D - l)$$
$$+ [8{\cdot}0603]\sin 2D + [7{\cdot}5715]\sin 2l$$
$$+ [7{\cdot}0112]\sin(2D - 2l) + [6{\cdot}1738]\sin(4D - 2l)$$
$$- [7{\cdot}3001]\sin 2F + [6{\cdot}4273]\sin(2D - 2F) - [6{\cdot}7818]\sin D$$
$$- [6{\cdot}7262]\sin(l + l') + [7{\cdot}0000]\sin(2D - l - l') + [6{\cdot}8555]\sin(l - l')$$
$$- [6{\cdot}3398]\sin(l + 2F) + [6{\cdot}2826]\sin(l - 2F)$$
$$+ [6{\cdot}2434]\sin 3l + [5{\cdot}8060]\sin(2D - 3l)$$
$$- [7{\cdot}5110]\sin l' + [6{\cdot}9040]\sin(2D - l'),$$

the coefficients being expressed in radians and the notation for the arguments being that of Delaunay. From these results the derivatives with respect to w_i are immediately obtainable with

$$D = w_1 - \text{earth's mean longitude},$$
$$l = w_1 - w_2,$$
$$F = w_1 - w_3.$$

For $\frac{dV}{dn}$ the results of Delaunay might be used, but it will be shown immediately that $\frac{dV}{dn}\delta n$ contributes nothing sensible to δV—a result probably of the use of the system n, c_2, c_3 instead of n, e, γ. The derivative has however been found for the largest terms in order to show this fact. The derivatives with respect to c_2, c_3 are obtained from my results with the help of equations (8). They give

$$n\frac{dV}{dn} = +[8{\cdot}0465]\sin l - [8{\cdot}4669]\sin(2\mathrm{D}-l) - [7{\cdot}2833]\sin(2\mathrm{D}+l)$$
$$- [8{\cdot}3965]\sin 2\mathrm{D} + [6{\cdot}8784]\sin 2l - [7{\cdot}1761]\sin(2\mathrm{D}-2l)$$
$$- [6{\cdot}8271]\sin 2\mathrm{F},$$
$$na^2\frac{dV}{dc_2} = -[11{\cdot}5828]\sin l - [10{\cdot}8895]\sin(2\mathrm{D}-l) - [9{\cdot}5959]\sin(2\mathrm{D}+l)$$
$$- [10{\cdot}0018]\sin 2\mathrm{D} - [10{\cdot}4147]\sin 2l - [9{\cdot}8546]\sin(2\mathrm{D}-2l),$$
$$na^2\frac{dV}{dc_3} = [9{\cdot}6959]\sin 2\mathrm{F}.$$

The latitude and parallax are treated in a similar manner.

Abbreviation of the Formulae for δV.

In the actual applications to the calculation of the direct inequalities much abbreviation of these formulae is possible. For example, in the large class which has its arguments independent of the lunar angles, $i_1 = i_2 = i_3 = 0$, and therefore $\delta n = \delta c_2 = \delta c_3 = 0$. The maximum values of the remaining coefficients are

$$f'\frac{A_1}{s} = 1''; \qquad \frac{f'}{s}\frac{dA}{de} = \frac{1}{5}''; \qquad \frac{f'}{s}\frac{dA}{d\mathrm{k}} = \frac{1}{10}'';$$
$$\text{in } \frac{dV}{dw_1} - 1,\ \frac{1}{9}; \qquad \text{in } \frac{dV}{dw_2},\ \frac{1}{9}; \qquad \text{in } \frac{dV}{dw_3},\ \frac{1}{250}.$$

Hence, to the accuracy desired, we can put

$$\delta w_1 = f'\frac{A_1}{s}\sin(qt+q'), \qquad \delta w_2 = [11{\cdot}0999]\frac{dA}{de}\frac{f'}{s}\sin(qt+q'),$$
$$\delta w_3 = [10{\cdot}2620]\frac{dA}{d\mathrm{k}}\frac{f'}{s}\sin(qt+q').$$

When i_1, i_2, i_3 are not all zero, it is still possible to limit the formulae very materially after δw_1 has been found, owing to the limitations on the magnitude of the coefficients in δw_1 and V.

For this purpose we note that, for the greatest coefficients,

$$n\left|\frac{dV}{dn}\right| < \frac{1}{30}, \qquad na^2\left|\frac{dV}{dc_2}\right| < 40, \qquad na^2\left|\frac{dV}{dc_3}\right| < \frac{1}{2};$$
$$\left|\frac{dV}{dw_1} - 1\right| < \frac{1}{9}, \qquad \left|\frac{dV}{dw_2}\right| < \frac{1}{9}, \qquad \left|\frac{dV}{dw_3}\right| < \frac{1}{250}.$$

Put
$$\lambda_1 = -i_1 + \cdot 01486 i_2 - \cdot 003744 i_3,$$
$$C = \lambda_1 \frac{fA}{s^2},$$
$$\frac{f'}{f} = [5\cdot 32390].$$

Then (11)
$$\begin{cases} \dfrac{\delta n}{n} = [5\cdot 3239]\, Cs \cos(qt + q'), \\ \dfrac{\delta c_2}{na^2} = [4\cdot 8419]\, \dfrac{Cs}{\lambda_1} i_2 \cos(qt + q'), \\ \dfrac{\delta c_3}{na^2} = [4\cdot 8419]\, \dfrac{Cs}{\lambda_1} i_3 \cos(qt + q'). \end{cases}$$

Also, if
$$\lambda_2 = \cdot 01486 i_1 - \cdot 007066 i_2 - \cdot 008148 i_3,$$
$$\lambda_3 = -\cdot 003744 i_1 - \cdot 008148 i_2 + \cdot 001210 i_3,$$
$$\mathrm{e}\frac{dA}{d\mathrm{e}} = j_2 A, \qquad \mathrm{k}\frac{dA}{d\mathrm{k}} = j_3 A,$$

the equations for δw_i become
$$\delta w_1 = C\left(1 + \frac{f's}{f\lambda_1} \cdot \frac{A_1}{A}\right) \sin(qt + q'),$$
$$\delta w_2 = \left\{-[3\cdot 4959]\frac{s}{\lambda_1} \cdot \frac{A_1}{A} + [7\cdot 3842]\frac{s}{\lambda_1} j_2 - [4\cdot 6150]\frac{s}{\lambda_1} j_3 + \frac{\lambda_2}{\lambda_1}\right\} C \sin(qt + q'),$$
$$\delta w_3 = \left\{+[2\cdot 8972]\frac{s}{\lambda_1} \cdot \frac{A_1}{A} + [5\cdot 9349]\frac{s}{\lambda_1} j_3 - [4\cdot 5805]\frac{s}{\lambda_1} j_2 + \frac{\lambda_3}{\lambda_1}\right\} C \sin(qt + q').$$

Now $\quad C < 15'', \quad \dfrac{f'A_1}{s} = C\dfrac{f's}{f\lambda_1} \cdot \dfrac{A_1}{A} < 0''\cdot 1, \quad \dfrac{A}{A_1} < 1, \quad \lambda_1 \gtreqless 2.$

Hence $\quad \dfrac{Cs}{\lambda_1} < 5000'', \quad Cs < 10000''.$

Therefore if the limits of the coefficients to be considered be $0''\cdot 003$, $\dfrac{\partial V}{\partial n}\delta n$ and the parts contributed by the first and third terms of δw_2 and δw_3 are insensible; as a matter of fact no one of these parts is so great as $0''\cdot 001$. Hence we obtain the final form of the equations of variations given on the next page. It is necessary to remember that, although the derivatives with respect to c_2, c_3 have been transformed into derivatives with respect to e, k, the derivative with respect to n is to be taken on the assumption that the coefficients are expressed in terms of n, c_2, c_3.

Final form of the abbreviated Equations of Variations.

$$(12)\quad \begin{cases} \delta w_1 = \left(\lambda_1 \dfrac{fA}{s^2} + \dfrac{f'A_1}{s}\right)\sin(qt+q') \\ \quad = C\left(1 + [5{\cdot}3239]\dfrac{s}{\lambda_1}\cdot\dfrac{A_1}{A}\right)\sin(qt+q'), \\ \delta w_2 = \left\{[7{\cdot}3840]\dfrac{s}{\lambda_1}j_2 + \dfrac{\lambda_2}{\lambda_1}\right\} C\sin(qt+q'), \\ \dfrac{\delta c_2}{na^2} = \left\{[4{\cdot}8419]\dfrac{s}{\lambda_1}i_2 C\right\}\cos(qt+q'), \\ \delta w_3 = \left\{[6{\cdot}9349]\dfrac{s}{\lambda_1}j_3 + \dfrac{\lambda_3}{\lambda_1}\right\} C\sin(qt+q'), \\ \dfrac{\delta c_3}{na^2} = \left\{[4{\cdot}8419]\dfrac{s}{\lambda_1}i_3 C\right\}\cos(qt+q'), \end{cases}$$

where
$$R = \frac{1}{4}\frac{m''}{m'}n'^2\mathrm{a}^2 A\cos(qt+q'),$$
$$\lambda_1 = -\,i_1 + {\cdot}01486 i_2 - {\cdot}003744 i_3,$$
$$\lambda_2 = +\,{\cdot}01486 i_1 - {\cdot}007066 i_2 - {\cdot}008148 i_3,$$
$$\lambda_3 = -\,{\cdot}003744 i_1 - {\cdot}008148 i_2 + {\cdot}001210 i_3,$$

i_1, i_2, i_3 are the coefficients of w_1, w_2, w_3 in $qt+q'$,

$$A_1 = \frac{n}{\mathrm{a}^2}\frac{d}{dn}(A\mathrm{a}^2), \qquad \mathrm{e}\frac{dA}{d\mathrm{e}} = j_2 A, \qquad \mathrm{k}\frac{dA}{d\mathrm{k}} = j_3 A.$$

One or two particular cases of frequent occurrence may be mentioned. In all cases $C = \lambda_1 \dfrac{fA}{s^2}$. For multiples of the arguments given, divide the second term in δw_1 by the multiple. The argument is the moon portion of $qt+q'$. The omitted terms are either zero or negligible.

Arg. 0. Here $i_1 = i_2 = i_3 = 0$.

$$\delta w_1 = \frac{f'A_1}{s}\sin(qt+q'),$$
$$\delta w_2 = [11{\cdot}0999]\,\frac{1}{A_1}\frac{dA}{d\mathrm{e}}\,\delta w_1,$$
$$\delta w_3 = [10{\cdot}2621]\,\frac{1}{A_1}\frac{dA}{d\mathrm{k}}\,\delta w_1.$$

Arg. $l = w_1 - w_2$.

$$\delta w_1 = \left\{1 - [5{\cdot}3175]\,s\frac{A_1}{A}\right\} C\sin(qt+q'),$$
$$\delta w_2 = \{-\,[7{\cdot}3776]\,s - {\cdot}02161\}\,C\sin(qt+q'),$$
$$\frac{\delta c_2}{na^2} = [4{\cdot}8355]\,Cs\cos(qt+q'),$$
$$\delta w_3 = -\,[6{\cdot}9285]\,j_3 Cs\sin(qt+q') \quad \text{(for the latitude only)}.$$

Arg. $2\mathrm{D} - l = w_1 + w_2 - 2T.$

$$\delta w_1 = \left\{1 \quad [5{\cdot}3304]\, s \frac{A_1}{A}\right\} C \sin(qt + q'),$$

$$\delta w_2 = -[7{\cdot}3905]\, sC \sin(qt + q'),$$

$$\frac{\delta c_2}{na^2} = -[4{\cdot}8484]\, sC \cos(qt + q').$$

Arg. $2l - 2\mathrm{D} = -2w_2 + 2T.$

$$\delta w_1 = \left\{1 - [6{\cdot}8505]\, s \frac{A_1}{A}\right\} C \sin(qt + q'),$$

$$\delta w_2 = \{-[9{\cdot}2119]\, s - {\cdot}4753\}\, C \sin(qt + q'),$$

$$\frac{\delta c_2}{na^2} = [6{\cdot}6698]\, sC \cos(qt + q').$$

There is quite an extensive class of terms containing this argument in which δw_1 is insensible. For these we can put

$$\delta w_2 = [5{\cdot}7642] \frac{f}{s} \cdot \frac{A}{\mathrm{e}^2} \sin(qt + q'),$$

$$\frac{\delta c_2}{na^2} = -[3{\cdot}2221] \frac{f}{s} \cdot \frac{A}{\mathrm{e}^2} \cos(qt + q').$$

The great majority of terms in R to be considered contain powers of the lunar eccentricity as a factor and the principal terms in the moon's true longitude have the same property. In nearly all such cases it is permissible to neglect higher powers of the lunar eccentricity and inclination. When these two conditions are satisfied it is not necessary to find δc_2, $\dfrac{dV}{dc_2}$. For then $j_2 = |i_2|$ and the ratio of the first term of the coefficient in δw_2 to that of δc_2 is the same as the ratio of the coefficients in $\mathrm{e}\dfrac{dV}{d\mathrm{e}}$, $\dfrac{dV}{dw_2}$. This arises from the fact that δw_2 depends mainly on $\mathrm{e}\dfrac{dR}{d\mathrm{e}}$, while δc_2 depends on $\dfrac{dR}{dw_2}$. If then

$$\delta w_2 = Q \sin(i_2 w_2 + \psi), \quad V = Q' \sin(i_2' w_2 + \psi'),$$

where ψ, ψ' are independent of w_2, and i_2, i_2' have the same sign,

$$\frac{dV}{dw_2} \delta w_2 + \frac{dV}{dc_2} \delta c_2 = QQ' i_2' \sin\{(i_2 - i_2')\, w_2 + \psi - \psi'\}.$$

It is to be noted that Q is here the first term in the coefficient of δw_2.

An exactly similar theorem holds with reference to c_3, w_3 and the terms which contain powers of k as a factor.

SECTION II.

THE TRANSFORMATION OF THE DISTURBING FUNCTION.

IT is well known that the disturbing function for the direct effect is R, where

$$\frac{R}{m''} = \frac{1}{[(\xi - x)^2 + (\eta - y)^2 + (\zeta - z)^2]^{\frac{1}{2}}} - \frac{x\xi + y\eta + z\zeta}{\Delta^3}.$$

In order to take into account the motion of the earth round its centre of mass, the terms containing $\frac{\mathrm{a}}{a'}$ are as usual to be multiplied by the ratio of the difference and sum of the masses of the earth and moon.

Put
$$x\frac{\partial}{\partial\xi} + y\frac{\partial}{\partial\eta} + z\frac{\partial}{\partial\zeta} = \frac{\partial}{\partial Q};$$

then, by Taylor's theorem,

$$(13) \quad \frac{1}{[(\xi - x)^2 + (\eta - y)^2 + (\zeta - z)^2]^{\frac{1}{2}}} = \left[1 - \frac{\partial}{\partial Q} + \frac{1}{\underline{|2}}\left(\frac{\partial}{\partial Q}\right)^2 - \frac{1}{\underline{|3}}\left(\frac{\partial}{\partial Q}\right)^3 + \dots\right]\frac{1}{\Delta}$$

$$= \frac{1}{\Delta} + \frac{x\xi + y\eta + z\zeta}{\Delta^3} + \left[\frac{1}{\underline{|2}}\left(\frac{\partial}{\partial Q}\right)^2 - \dots\right]\frac{1}{\Delta}.$$

Since R will enter only through its derivatives with respect to the lunar elements which are not present in Δ, we obtain

$$(14) \quad \frac{R}{m''} = \left[\frac{1}{\underline{|2}}\left(\frac{\partial}{\partial Q}\right)^2 - \frac{1}{\underline{|3}}\left(\frac{\partial}{\partial Q}\right)^3 + \dots\right]\frac{1}{\Delta}.$$

The separation of the terms of R into sums of products, one factor involving the lunar coordinates and the other the planet's coordinates, suggested by Hill* and adopted by Radau†, is implicitly used here, and the division is the same in principle, but it will take a different form with the use of complex coordinates, and the method of separation is made on a general plan which can be applied to terms of any order in R.

* *Amer. Eph. Papers*, Vol. III.

† *Ann. de l'Obs. de Paris* (Mém.), Vol. XXI., 1892.

Put
$$x + y\sqrt{-1} = \mathrm{u}, \qquad \xi + \eta\sqrt{-1} = \mathrm{u}_1,$$
$$x - y\sqrt{-1} = \mathrm{s}, \qquad \xi - \eta\sqrt{-1} = \mathrm{s}_1;$$
then
$$\Delta^2 = \mathrm{u}_1\mathrm{s}_1 + \zeta^2, \qquad r^2 = \mathrm{us} + z^2,$$
$$\frac{\partial}{\partial Q} = \mathrm{u}\frac{\partial}{\partial \mathrm{u}_1} + \mathrm{s}\frac{\partial}{\partial \mathrm{s}_1} + z\frac{\partial}{\partial \zeta}.$$

And since $1/\Delta$ is a solution of Laplace's equation with respect to ξ, η, ζ,
$$\frac{\partial^2}{\partial \zeta^2}\frac{1}{\Delta} = -\left(\frac{\partial^2}{\partial \xi^2} + \frac{\partial^2}{\partial \eta^2}\right)\frac{1}{\Delta} = -4\frac{\partial^2}{\partial \mathrm{u}_1 \partial \mathrm{s}_1}\frac{1}{\Delta}.$$

Hence, omitting $1/\Delta$ for the sake of brevity,
$$\frac{1}{2}\left(\frac{\partial}{\partial Q}\right)^2 = \frac{1}{2}\left(\mathrm{u}\frac{\partial}{\partial \mathrm{u}_1}\right)^2 + \frac{1}{2}\left(\mathrm{s}\frac{\partial}{\partial \mathrm{s}_1}\right)^2 + \mathrm{us}\frac{\partial^2}{\partial \mathrm{u}_1 \partial \mathrm{s}_1}$$
$$+ \frac{1}{2}z^2\left(\frac{\partial}{\partial \zeta}\right)^2 + z\frac{\partial}{\partial \zeta}\left(\mathrm{u}\frac{\partial}{\partial \mathrm{u}_1} + \mathrm{s}\frac{\partial}{\partial \mathrm{s}_1}\right)$$
$$= \text{real part of } (r^2 - 3z^2)\frac{\partial^2}{\partial \mathrm{u}_1 \partial \mathrm{s}_1} + \mathrm{u}^2\left(\frac{\partial}{\partial \mathrm{u}_1}\right)^2 + 2z\mathrm{u}\frac{\partial^2}{\partial \zeta \partial \mathrm{u}_1},$$
$$-\frac{1}{\underline{|3}}\left(\frac{\partial}{\partial Q}\right)^3 = \text{real part of } -(r^2 - 5z^2)\,\mathrm{u}\frac{\partial^3}{\partial \mathrm{u}_1^2 \partial \mathrm{s}_1} - \frac{1}{3}\mathrm{u}^3\left(\frac{\partial}{\partial \mathrm{u}_1}\right)^3$$
$$- z\mathrm{u}^2\frac{\partial^3}{\partial \zeta \partial \mathrm{u}_1^2} - z\left(r^2 - \frac{5}{3}z^2\right)\frac{\partial^3}{\partial \zeta \partial \mathrm{u}_1 \partial \mathrm{s}_1},$$
and so on. The method is the same throughout: expand $\left(\frac{\partial}{\partial Q}\right)^n$ and replace $\left(\frac{\partial}{\partial \zeta}\right)^{2m}$ by $\left(-4\frac{\partial^2}{\partial \mathrm{u}_1 \partial \mathrm{s}_1}\right)^m$ and us by $r^2 - z^2$.

I now give a method for replacing the derivatives with respect to the coordinates, by derivatives with respect to a', T (the earth's mean longitude), h'', γ''. In this way it is possible to make use of Leverrier's expansion* of $1/\Delta$ in terms of the elliptic elements of two planets to the seventh order inclusive with respect to their eccentricities and their mutual inclination.

Expressing the coordinates in polar coordinates, we have
$$(15)\quad \begin{cases} \xi = -r' + (1 - \gamma''^2)\, r'' \cos(V'' - V') + \gamma''^2 r'' \cos(V'' + V' - 2h''), \\ \eta = \qquad (1 - \gamma''^2)\, r'' \sin(V'' - V') - \gamma''^2 r'' \sin(V'' + V' - 2h''), \\ \zeta = \quad 2\gamma''\sqrt{1 - \gamma''^2}\, r'' \sin(V'' - h''). \end{cases}$$
Whence
$$\Delta^2 = r''^2 - r'^2 - 2r'\xi,$$
$$\mathrm{u}_1 = -r' + (1 - \gamma''^2)\exp.(V'' - V')\sqrt{-1}$$
$$+ \gamma''^2 r'' \exp.(-V'' - V' + 2h'')\sqrt{-1},$$
the expression for s_1 being obtained by changing the sign of $\sqrt{-1}$.

* See below.

From these results we obtain

$$(16)\quad \begin{cases} \dfrac{\partial}{\partial r'} = -\dfrac{\partial}{\partial \mathrm{u}_1} - \dfrac{\partial}{\partial \mathrm{s}_1}, & \sqrt{-1}\,\dfrac{\partial}{\partial V'} = (\mathrm{u}_1 + r')\dfrac{\partial}{\partial \mathrm{u}_1} - (\mathrm{s}_1 + r')\dfrac{\partial}{\partial \mathrm{s}_1}, \\ \dfrac{\partial}{\partial \mathrm{v}} = r'\dfrac{\partial}{\partial r'} + \sqrt{-1}\,\dfrac{\partial}{\partial V'} = & \mathrm{u}_1\dfrac{\partial}{\partial \mathrm{u}_1} - \mathrm{s}_1\dfrac{\partial}{\partial \mathrm{s}_1} - 2r'\dfrac{\partial}{\partial \mathrm{s}_1}, \\ \dfrac{\partial}{\partial \mathrm{w}} = r'\dfrac{\partial}{\partial r'} - \sqrt{-1}\,\dfrac{\partial}{\partial V'} = & -\mathrm{u}_1\dfrac{\partial}{\partial \mathrm{u}_1} + \mathrm{s}_1\dfrac{\partial}{\partial \mathrm{s}_1} - 2r'\dfrac{\partial}{\partial \mathrm{u}_1}, \end{cases}$$

the notation $\dfrac{\partial}{\partial \mathrm{v}}, \dfrac{\partial}{\partial \mathrm{w}}$ being introduced for a few pages for brevity.

Let F denote a function of ζ and of the product $\mathrm{u}_1\mathrm{s}_1$ only. Then

$$(16a)\qquad \frac{\partial F}{\partial \mathrm{v}} = -2r'\frac{\partial F}{\partial \mathrm{s}_1}, \qquad \frac{\partial F}{\partial \mathrm{w}} = -2r'\frac{\partial F}{\partial \mathrm{u}_1},$$

$$(17)\qquad \left(\frac{\partial}{\partial \mathrm{u}_1}\right)^n \frac{1}{\Delta} = \mathrm{s}_1{}^n F, \qquad \left(\frac{\partial}{\partial \mathrm{s}_1}\right)^n \frac{1}{\Delta} = \mathrm{u}_1{}^n F.$$

Hence $$\frac{\partial^2}{\partial \mathrm{v}\,\partial \mathrm{w}} \cdot \frac{1}{\Delta} = \left[\left(r'\frac{\partial}{\partial r'}\right)^2 + \left(\frac{\partial}{\partial V'}\right)^2\right]\frac{1}{\Delta} = \frac{\partial}{\partial \mathrm{v}}\left(-2r'\frac{\partial}{\partial \mathrm{u}_1} \cdot \frac{1}{\Delta}\right)$$
$$= 4r'^2 \frac{\partial^2}{\partial \mathrm{u}_1 \partial \mathrm{s}_1} \cdot \frac{1}{\Delta},$$

$$\left(\frac{\partial}{\partial \mathrm{v}} - 2\right)\frac{\partial}{\partial \mathrm{v}} \cdot \frac{1}{\Delta} = \left(\frac{\partial}{\partial \mathrm{v}} - 2\right)\left(-2r'\frac{\partial}{\partial \mathrm{s}_1} \cdot \frac{1}{\Delta}\right) = -2r'\left(\frac{\partial}{\partial \mathrm{v}} - 1\right)\left(\frac{\partial}{\partial \mathrm{s}_1} \cdot \frac{1}{\Delta}\right)$$
$$= 4r'^2 \frac{\partial^2}{\partial \mathrm{s}_1{}^2}\frac{1}{\Delta}.$$

Similarly,

$$\left(\frac{\partial}{\partial \mathrm{v}} - 4\right)\left(\frac{\partial}{\partial \mathrm{v}} - 2\right)\frac{\partial}{\partial \mathrm{v}} \cdot \frac{1}{\Delta} = (-2r')^3 \frac{\partial^3}{\partial \mathrm{s}_1{}^3}\frac{1}{\Delta},$$

$$\left(\frac{\partial}{\partial \mathrm{v}} - 2\right)\frac{\partial}{\partial \mathrm{v}} \cdot \frac{\partial}{\partial \mathrm{w}} \cdot \frac{1}{\Delta} = (-2r')^3 \frac{\partial^3}{\partial \mathrm{u}_1 \partial \mathrm{s}_1{}^2}\frac{1}{\Delta},$$

and the general law is evident. The derivatives with respect to u_1, s_1 are thus expressed in terms of derivatives with respect to r', V', when for $\dfrac{\partial}{\partial \mathrm{v}}, \dfrac{\partial}{\partial \mathrm{w}}$ have been substituted their values (16).

Next,

$$\Delta^2 = r'^2 + r''^2 - 2r'r''(1 - \gamma''^2)\cos(V'' - V') - 2r'r''\gamma''^2\cos(V'' + V' - 2h'')$$
$$= r'^2 + r''^2 - 2r'r''\cos(V'' - V') + 4r'r''\gamma''^2\sin(V'' - h'')\sin(V' - h'').$$

Hence $$\left(\frac{\partial}{\partial V'} + \frac{1}{2\gamma''^2} \cdot \frac{\partial}{\partial h''}\right)\frac{1}{\Delta} = \frac{2(1 - \gamma''^2)\,r'r''\sin(V'' - h'')\cos(V' - h'')}{\Delta^3},$$

$$(1 - \gamma''^2)\frac{\partial}{\partial \gamma''^2}\frac{1}{\Delta} = -\frac{2(1 - \gamma''^2)\,r'r''\sin(V'' - h'')\sin(V' - h'')}{\Delta^3}.$$

Therefore

$$\left[\frac{\partial}{\partial V'} + \frac{1}{2\gamma''^2} \cdot \frac{\partial}{\partial h''} + \sqrt{-1}\,(1-\gamma''^2)\frac{\partial}{\partial \gamma''^2}\right]\frac{1}{\Delta}$$

$$= \frac{2(1-\gamma''^2)\, r'r'' \sin(V''-h'') \exp. -\sqrt{-1}\,(V'-h'')}{\Delta^3}$$

$$= \frac{\sqrt{1-\gamma''^2}}{\gamma''}\frac{r'\zeta}{\Delta^3} \exp. -\sqrt{-1}\,(V'-h'').$$

This gives

$$\frac{\partial}{\partial \zeta} \cdot \frac{1}{\Delta} = -\frac{\zeta}{\Delta^3} = -\frac{\gamma''}{\sqrt{1-\gamma''^2}} \frac{\exp. \sqrt{-1}\,(V'-h'')}{r'}$$
$$\left\{\frac{\partial}{\partial V'} + \frac{1}{2\gamma''^2}\frac{\partial}{\partial h''} + \sqrt{-1}\,(1-\gamma''^2)\frac{\partial}{\partial \gamma''^2}\right\}\frac{1}{\Delta}.$$

(As ζ/Δ^3 is real, we incidentally have the relation

$$\left[\sin(V'-h'') \cdot \left(\frac{\partial}{\partial V'} + \frac{1}{2\gamma''^2}\frac{\partial}{\partial h''}\right) - \cos(V'-h'')(1-\gamma''^2)\frac{\partial}{\partial \gamma''^2}\right]\frac{1}{\Delta} = 0,$$

and the sign of the $\sqrt{-1}$ may be changed.)

Since ζ/Δ^3 involves u_1, s_1 in their product only, the formulae (16a) are available; and as

$$\frac{\partial}{\partial v}\frac{\exp. \sqrt{-1}\,(V'-h'')}{r'} F = \frac{\exp. +\sqrt{-1}\,(V'-h'')}{r'}\left(\frac{\partial}{\partial v} - 2\right) F,$$

$$\frac{\partial}{\partial w}\frac{\exp. \sqrt{-1}\,(V'-h'')}{r'} F = \frac{\exp. +\sqrt{-1}\,(V'-h'')}{r'}\frac{\partial}{\partial w} F,$$

we obtain, by substitution of the various results,

$$(18)\quad R = \frac{m''}{4r'^2}\left[(r^2 - 3z^2)\frac{\partial^2}{\partial v \partial w} + u^2 \frac{\partial}{\partial w}\left(\frac{\partial}{\partial w} - 2\right)\right.$$
$$+ 4zu\frac{\gamma''}{\sqrt{1-\gamma''^2}} \exp. \sqrt{-1}\,(V'-h'')\frac{\partial}{\partial w}\left(\frac{\partial}{\partial V'} + \frac{1}{2\gamma''^2}\frac{\partial}{\partial h''}\right.$$
$$\left.\left. + \sqrt{-1}\,(1-\gamma''^2)\frac{\partial}{\partial \gamma''^2}\right]\frac{1}{\Delta}\right.$$

$$+ \frac{m''}{8r'^3}\left[u(r^2 - 5z^2)\frac{\partial^2}{\partial v \partial w}\left(\frac{\partial}{\partial w} - 2\right)\right.$$
$$\left. + \frac{1}{3}u^3 \frac{\partial}{\partial w}\left(\frac{\partial}{\partial w} - 2\right)\left(\frac{\partial}{\partial w} - 4\right)\right]\frac{1}{\Delta}$$

$$+ \frac{m''}{4r'^3}\frac{\gamma''}{\sqrt{1-\gamma''^2}} \exp. \sqrt{-1}\,(V'-h'')\left[z\left(r^2 - \frac{5}{3}z^2\right)\left(\frac{\partial}{\partial v} - 2\right)\frac{\partial}{\partial w}\right.$$
$$\left. + zu^2\left(\frac{\partial}{\partial w} - 2\right)\frac{\partial}{\partial w}\right]\left[\frac{\partial}{\partial V'} + \frac{1}{2\gamma''^2}\frac{\partial}{\partial h''} + \sqrt{-1}\,(1-\gamma''^2)\frac{\partial}{\partial \gamma''^2}\right]\frac{1}{\Delta}.$$

The derivatives with respect to v, w are easily expressed in terms of derivatives with respect to the elements by means of the well known formulae

$$r' \frac{\partial}{\partial r'} = a' \frac{\partial}{\partial a'}, \quad \frac{\partial}{\partial V'} = \frac{\partial}{\partial T},$$

where T is the mean longitude of the earth and where it is supposed that the angles referring to the earth are expressed in terms of T, l'. Also, Leverrier's development depends on $\alpha = \frac{a_1}{a_2}$, where a_1 is the mean distance of the inner planet, a_2 that of the outer. Hence if we put

$$\alpha \frac{\partial}{\partial \alpha} = D$$

we have, since $\frac{1}{\Delta}$ is a homogeneous function of degree -1 with respect to a', a'',

$$a_1 = a'', \quad a_2 = a', \quad r' \frac{\partial}{\partial r'} = -\alpha \frac{\partial}{\partial \alpha} - 1 = -D - 1 \quad \text{for inner planets,}$$

$$a_1 = a', \quad a_2 = a'', \quad r' \frac{\partial}{\partial r'} = \alpha \frac{\partial}{\partial \alpha} = D \quad \text{for outer planets.}$$

The planet portions of the disturbing function are therefore expressed in terms of derivatives of $1/\Delta$ *with respect to* $\alpha, T, h'', \gamma''^2$,*—the quantities present explicitly in Leverrier's development.*

Each term in the development of R consists of two main parts together with a constant factor. The first part is a function of the moon's coordinates alone $\left(\text{with which will be combined the factors } \frac{1}{r'^2}, \frac{1}{r'^3}, e^{(V'-h'')\sqrt{-1}}\right)$; the second part consists of the derivatives of $1/\Delta$ which is a function of the earth's and of the planet's elements only. Let θ denote an angle present in the first part, ϕ an angle present in the second part, that is, in $1/\Delta$, so that $\theta \pm \phi$ is the argument of a term in R which is under consideration. Put j for $\sqrt{-1}$, and*

$$\frac{a'^2}{r'^2} \cdot \frac{r^2 - 3z^2}{\mathrm{a}^2} = M_1 (e^{j\theta} + e^{-j\theta}) = 2M_1 \cos \theta,$$

$$\frac{a'^2}{r'^2} \cdot \frac{\mathrm{u}^2}{\mathrm{a}^2} = \tfrac{1}{2}(M_2 + M_3) e^{j\theta} + \tfrac{1}{2}(M_2 - M_3) e^{-j\theta} = M_2 \cos \theta + jM_3 \sin \theta,$$

$$\frac{a'^2}{r'^2} \frac{zj \,.\, \mathrm{u} e^{j(V'-h'')}}{\mathrm{a}^2} = M_4 e^{j\theta} \dagger = M_4 \cos \theta + jM_4 \sin \theta.$$

These functions are calculated once for all and serve for all the planets.

* The exponent always attached here to the exponential e prevents any confusion with the same symbol used to represent the lunar eccentricity as defined by Delaunay.

† Owing to the absence of h'' from z, u, r', the part $e^{-j\theta}$ is not present here.

For the second part put*

$$\frac{1}{\Delta} = P\cos\phi = P\cos(iT + 2i_1h'' + \phi'),$$

where ϕ is independent of T, h''. Also for *inner* planets

$$P_1\cos\phi = \frac{\partial^2}{\partial \mathrm{v}\,\partial \mathrm{w}}\cdot\frac{1}{\Delta} = \left[\left(r'\frac{\partial}{\partial r'}\right)^2 + \frac{\partial}{\partial V'^2}\right]\frac{1}{\Delta} = [(D+1)^2 - i^2]\,P\cos\phi,$$

$$2P_2\cos\phi - 2P_3 j\sin\phi = \frac{\partial}{\partial \mathrm{w}}\left(\frac{\partial}{\partial \mathrm{w}} - 2\right)\frac{1}{\Delta}$$

$$= \left(D + 1 + j\frac{\partial}{\partial V'}\right)\left(D + 3 + j\frac{\partial}{\partial V'}\right)P\cos\phi,$$

$$P_4\cos\phi + jP_5\sin\phi = -\left(D + 1 + j\frac{\partial}{\partial V'}\right)$$

$$\left[\frac{\partial}{\partial V'} + \frac{1}{2\gamma''^2}\frac{\partial}{\partial h''} + j(1-\gamma''^2)\frac{\partial}{\partial \gamma''^2}\right]jP\cos\phi;$$

so that

$$P_1 = [(D+1)^2 - i^2]\,P,$$

$$P_2 = \tfrac{1}{2}[(D+1)(D+3) + i^2]\,P = \tfrac{1}{2}P_1 + (D+2)\,P + (i^2-1)\,P,$$

$$P_3 = (D+2)\,iP,$$

$$P_4 + P_5 = (D+1-i)\left\{\ i + \frac{i_1}{\gamma''^2} + (1-\gamma''^2)\frac{\partial}{\partial \gamma''^2}\right\}P,$$

$$P_4 - P_5 = (D+1+i)\left\{-i - \frac{i_1}{\gamma''^2} + (1-\gamma''^2)\frac{\partial}{\partial \gamma''^2}\right\}P.$$

The first line of R, which consists of those terms in (18) which have the factor $\frac{m''}{4r'^2}$, becomes with these substitutions, after inserting the unit factor† $\frac{n'^2a'^3}{m'}$,

$$(19)\quad R_1 = \frac{m''}{4m'}n'^2\mathrm{a}^2a'\left[M_1P_1 + M_2P_2 \mp M_3P_3\right.$$

$$\left. - \frac{2\gamma''}{\sqrt{1-\gamma''^2}}M_4(P_4 \pm P_5)\right]\cos(\theta \pm \phi),$$

either all the upper signs or all the lower signs being taken according as it is convenient to use the sum or difference of the angles θ, ϕ. It will be noticed that the P_p contain the divisor a' which will cancel the a' outside the parentheses; otherwise they contain a', a'' only in the form $a'/a'' = \alpha$.

* The letter i has also been used elsewhere in a different connection; but no confusion need occur.

† The denominator is properly the sum of the masses of the earth, moon and sun; but the difference is insensible.

For exterior planets the same formulae hold if (1) we substitute $-D-1$ for D, (2) replace the factor a' by $\alpha a''$, since Leverrier's formulae are then to be expressed in the form $\frac{1}{a''}$ func. (α).

For the second part of R, which involves the factor $\frac{\mathrm{a}}{a'}$ and which will be denoted by R_2, put for the moon portions,

$$\frac{a'^3}{r'^3}\frac{\mathrm{u}\,(r^2-5z^2)}{\mathrm{a}^3} = \tfrac{1}{2}(M_6+M_7)\,e^{\theta j} + \tfrac{1}{2}(M_6-M_7)\,e^{-\theta j} = M_6\cos\theta + jM_7\sin\theta,$$

$$\frac{1}{3}\frac{a'^3}{r'^3}\cdot\frac{\mathrm{u}^3}{\mathrm{a}^3} = \tfrac{1}{2}(M_8+M_9)\,e^{\theta j} + \tfrac{1}{2}(M_8-M_9)\,e^{-\theta j} = M_8\cos\theta + jM_9\sin\theta,$$

$$\frac{a'^3}{r'^3}\cdot\frac{jz\,.\,e^{j(V'-h'')}}{\mathrm{a}^3}\,(r^2-\tfrac{5}{3}z^2) = M_{10}e^{\theta j} = M_{10}(\cos\theta + j\sin\theta),$$

$$\frac{a'^3}{r'^3}\cdot\frac{jz\,.\,e^{j(V'-h'')}\,\mathrm{u}^2}{\mathrm{a}^3} = M_{12}e^{\theta j} = M_{12}(\cos\theta + j\sin\theta),$$

and for the planet portions, in the case of the inner planets,

$$P_6\cos\phi - jP_7\sin\phi = \left(\frac{\partial}{\partial\mathrm{w}} - 2\right)\frac{\partial^2}{\partial\mathrm{v}\,\partial\mathrm{w}}\,.\,P\cos\phi$$

$$= -\left(D+3+j\frac{\partial}{\partial V'}\right)\left[(D+1)^2 - i^2\right]P\cos\phi,$$

$$P_8\cos\phi - jP_9\sin\phi = \left(\frac{\partial}{\partial\mathrm{w}} - 4\right)\left(\frac{\partial}{\partial\mathrm{w}} - 2\right)\frac{\partial}{\partial\mathrm{w}}\,P\cos\phi$$

$$= -\left(D+5+j\frac{\partial}{\partial V'}\right)\left(D+3+j\frac{\partial}{\partial V'}\right)\left(D+1+j\frac{\partial}{\partial V'}\right)P\cos\phi,$$

$$P_{10}\cos\phi + jP_{11}\sin\phi = j\left(\frac{\partial}{\partial\mathrm{v}} - 2\right)\frac{\partial}{\partial\mathrm{w}}\left\{\frac{\partial}{\partial V'} + \frac{1}{2\gamma''^2}\frac{\partial}{\partial h''} + j(1-\gamma''^2)\frac{\partial}{\partial\gamma''^2}\right\}P\cos\phi$$

$$= \left(D+3-j\frac{\partial}{\partial V'}\right)\left(D+1+j\frac{\partial}{\partial V'}\right)\left\{-(1-\gamma''^2)\frac{\partial}{\partial\gamma''^2} + j\left(\frac{\partial}{\partial V'} + \frac{1}{2\gamma''^2}\frac{\partial}{\partial h''}\right)\right\}P\cos\phi,$$

$$P_{12}\cos\phi + jP_{13}\sin\phi = j\left(\frac{\partial}{\partial\mathrm{w}} - 2\right)\frac{\partial}{\partial\mathrm{w}}\left\{\frac{\partial}{\partial V'} + \frac{1}{2\gamma''^2} + j(1-\gamma''^2)\frac{\partial}{\partial\gamma''^2}\right\}P\cos\phi$$

$$= \left(D+3+j\frac{\partial}{\partial V'}\right)\left(D+1+j\frac{\partial}{\partial V'}\right)\left\{-(1-\gamma''^2)\frac{\partial}{\partial\gamma''^2} + j\left(\frac{\partial}{\partial V'} + \frac{1}{2\gamma''^2}\frac{\partial}{\partial h''}\right)\right\}P\cos\phi,$$

so that

$$P_6 \pm P_7 = -\{(D+1)^2 - i^2\}(D + 3 \pm i) P,$$

$$P_8 = -(D^2 + 6D + \text{[illegible]} + \text{[illegible]})(D + \text{[illegible]}) P,$$

$$P_9 = -(3D^2 + 18D + 23 + i^2) iP,$$

$$P_{10} \pm P_{11} = -\{(D+3)(D+1) - i^2 \mp 2i\}\left\{(1-\gamma''^2)\frac{\partial}{\partial \gamma''^2} \pm \left(i + \frac{i_1}{\gamma''^2}\right)\right\} P,$$

$$P_{12} \pm P_{13} = -\{(D+3)(D+1) + i^2 \mp 2i(D+2)\}\left\{(1-\gamma''^2)\frac{\partial}{\partial \gamma''^2}\left(i + \frac{i_1}{\gamma''^2}\right)\right\} P.$$

Then

$$(20) \quad R_2 = \frac{1}{16}\frac{m''}{m'} \cdot n'^2 a^2 \frac{a}{a'} a' \Big[M_6 P_6 \mp M_7 P_7 + M_8 P_8 \mp M_9 P_9$$

$$- \frac{2\gamma''}{\sqrt{1-\gamma''^2}} \{M_{10}(P_{10} \pm P_{11}) + M_{12}(P_{12} \pm P_{13})\}\Big] \cos(\theta \pm \phi).$$

The changes for outer planets are the same as in the case of R_1.

The moon coordinate u is referred to the *true* place of the sun. As my results will be used to calculate the functions of u, s, z, and as these results are referred to the *mean* place of the sun, it is necessary to replace u by $u_0 e^{-j(V'-T)}$ where u_0 is the same as the u of my Lunar Theory.

The formulae for M_p and P_p are collected and slightly altered in form at the end of this Section. In appearance they are somewhat complicated, but they are easy and rapid for numerical calculation. For example, the effect of the first factor in P_4, P_5, P_{10}, P_{11}, P_{12}, P_{13} on the expansion of $1/\Delta$ is seen at a glance; it is shown below that where P_1 has not been tabulated it is rapidly found from a general formula; and the effect of D on P or P_1 requires nothing more than the replacing of K_p by $(p+1)K_{p+1} + pK_p$. The great majority of terms involve only P_1, P_2, P_3, while the functions P_6, P_7, ... are required only for the very few sensible terms which have the factor $\frac{a}{a'}$.

In the above development it is assumed that the elliptic values of ξ, η, ζ are used. If the effect of their perturbations is to be included, the new part δR of R can be found by operating on its value (14) with

$$\delta\xi \frac{\partial}{\partial \xi} + \delta\eta \frac{\partial}{\partial \eta} + \delta\zeta \frac{\partial}{\partial \zeta},$$

where $\delta\xi$, $\delta\eta$, $\delta\zeta$ are the disturbed values (supposed known) of ξ, η, ζ.

Similarly, x, y, z represent the coordinates of the moon as disturbed by a sun moving in a fixed elliptic orbit. If δx, δy, δz are the perturbations produced in x, y, z by the value of R in (18), we can include the effect of these additions to x, y, z (perturbations of the second order relatively to the masses) by a further disturbing function

$$\delta' R = \delta x \frac{\partial R}{\partial x} + \delta y \frac{\partial R}{\partial y} + \delta z \frac{\partial R}{\partial z};$$

but then the w_i, c_i must be newly defined, that is, they require slight additions—probably insensible—in the equations of variations.

There is no difficulty in the calculation of these two new functions. They are obtained easily by the method given above. It will be found that δR, for the first term of R, involves the functions $P_6, \ldots, P_{13}$ instead of $P_1, \ldots, P_5$, while the expansion of $\delta' R$, which involves changes in the M_p only, is so simple that its value is obtained immediately.

The formulae to be used are collected below.

The Functions $M_1, M_2, \ldots$.

$$M_1 = \text{coef. of } e^{j\theta} \quad \text{in} \quad \frac{a'^2}{r'^2} \cdot \frac{r^2 - 3z^2}{\mathrm{a}^2},$$

$$\tfrac{1}{2}(M_2 \pm M_3) = \text{coef. of } e^{\pm j\theta} \quad \text{in} \quad \frac{a'^2}{r'^2} e^{\mp 2j(V'-T)} \cdot \frac{\mathrm{u}_0^2}{\mathrm{a}^2},$$

$$M_4 = \text{coef. of } e^{j\theta} \quad \text{in} \quad \frac{a'^2}{r'^2} e^{j(T-h'')} \cdot \frac{\mathrm{u}_0 zj}{\mathrm{a}^2},$$

$$\tfrac{1}{2}(M_6 \pm M_7) = \text{coef. of } e^{\pm j\theta} \quad \text{in} \quad \frac{a'^3}{r'^3} e^{\mp j(V'-T)} \cdot \frac{\mathrm{u}_0 (r^2 - 3z^2)}{\mathrm{a}^3},$$

$$\tfrac{1}{2}(M_8 \pm M_9) = \text{coef. of } e^{\pm j\theta} \quad \text{in} \quad \frac{a'^3}{r'^3} e^{\mp 3j(V'-T)} \cdot \frac{\mathrm{u}_0^3}{3\mathrm{a}^3},$$

$$M_{10} = \text{coef. of } e^{j\theta} \quad \text{in} \quad \frac{a'^3}{r'^3} e^{j(V'-h'')} \cdot \frac{zj(r^2 - \frac{5}{3}z^2)}{\mathrm{a}^3},$$

$$M_{12} = \text{coef. of } e^{j\theta} \quad \text{in} \quad \frac{a'^3}{r'^3} e^{-j(V'+h''-2T)} \cdot \frac{zj\mathrm{u}_0^2}{\mathrm{a}^3}.$$

The Functions $P_1, P_2, \ldots$.

$$\frac{1}{\Delta} = \Sigma P \cos\phi = \Sigma P \cos(iT + 2i_1 h'' + \phi'), \qquad D = \alpha \frac{d}{d\alpha}.$$

Inner Planets.

$$P_1 = \{(D+1)^2 - i^2\} P,$$

$$P_2 = \tfrac{1}{2} P_1 + (D+2) P + (i^2 - 1) P,$$

$$P_3 = i(D+2) P,$$

$$P_4 \pm P_5 = \left\{(1-\gamma''^2)\frac{\partial}{\partial \gamma''^2} \pm \left(i + \frac{i_1}{\gamma''^2}\right)\right\}(D + 1 \mp i) P,$$

$$P_6 \pm P_7 = -(D + 3 \pm i) P_1,$$

$$P_8 \pm P_9 = -(D \pm 3i + 7) P_1 - 4(1 \pm i)(2 \pm i)(D \pm i + 1) P,$$

$$P_{10} \pm P_{11} = -\left[(1-\gamma''^2)\frac{\partial}{\partial \gamma''^2} \pm \left(i + \frac{i_1}{\gamma''^2}\right)\right][P_1 + 2DP + (2 \mp 2i) P],$$

$$P_{12} \pm P_{13} = -\left[(1-\gamma''^2)\frac{\partial}{\partial \gamma''^2} \pm \left(i + \frac{i_1}{\gamma''^2}\right)\right][P_1 + (2 \mp 2i) DP + 2(i \mp 1)^2 P].$$

Outer Planets.

$$P_1 = (D^2 - i^2)\,P,$$

$$P_2 = \tfrac{1}{2}P_1 - DP\ [illegible]\ P,$$

$$P_3 = -i\,(D-1)\,P,$$

$$P_4 \pm P_5 = -(D \pm i)\left\{(1-\gamma''^2)\frac{\partial}{\partial\gamma''^2} \pm \left(i + \frac{i_1}{\gamma''^2}\right)\right\} P,$$

$$P_6 \pm P_7 = (D - 2 \mp i)\,P_1,$$

$$P_8 \pm P_9 = (D \mp 3i - 6)\,P_1 + 4\,(1 \pm i)\,(2 \pm i)\,(D \mp i)\,P,$$

$$P_{10} \pm P_{11} = \left[(1-\gamma''^2)\frac{\partial}{\partial\gamma''^2} \pm \left(i + \frac{i_1}{\gamma''^2}\right)\right][-P_1 + 2DP \pm 2iP],$$

$$P_{12} \pm P_{13} = -\left[(1-\gamma''^2)\frac{\partial}{\partial\gamma''^2} \pm \left(i + \frac{i_1}{\gamma''^2}\right)\right][P_1 + (\pm 2i - 2)\,DP + (2i^2 \mp 2i)\,P].$$

The Disturbing Function is $R_1 + R_2$ where R_1 has the value given by (19) on p. 21 and R_2 the value given by (20) on p. 23.

Owing to the fact that M_2 is nearly equal to $\pm M_3$ and P_2 to $\mp P_3$ in most cases, the portion $M_2P_2 \mp M_3P_3$ of R_1 was usually put into the form

$$\tfrac{1}{2}(M_2 + M_3)(P_2 \mp P_3) + \tfrac{1}{2}(M_2 - M_3)(P_2 \pm P_3)$$

for computation; the degree of accuracy of the tables in Section V. was then always sufficient.

SECTION III.

DEVELOPMENT OF THE DISTURBING FUNCTION.

Leverrier's Expansion of $\frac{1}{\Delta}$.

LEVERRIER* expands in powers of the eccentricities of the two planets and the square of the mutual inclination of their orbits. The arguments are their mean motions, the longitudes of their perigees measured along the fixed plane (ecliptic at a given date) to the nodes, then along the orbits, and the longitudes of the nodes. As one of the planets in the present investigation is taken to be the earth, and the fixed plane the ecliptic, supposed immovable, Leverrier's arguments become in the present notation, for an inner planet,

$$\tau = \tau' = h'', \qquad l' = T, \qquad \omega = T - l',$$
$$\lambda = P, \qquad \varpi' = P - l''.$$

(Leverrier denotes the mean longitude of the earth by l'; I use this letter for the mean anomaly.)

For the coefficients I use the same notation with the exception of η, which I denote by γ''. For an outer planet the accents (except in the case of h'', γ'') interchange.

Put
$$\Delta_0^2 = 1 + \alpha^2 - 2\alpha\cos(T - P),$$
then with Leverrier, using Σ to denote summation for integral values of i from $+\infty$ to $-\infty$, I put

$$\frac{1}{\Delta_0} = \tfrac{1}{2}a'\Sigma A^{(i)}\cos i(T - P),$$

$$\frac{\alpha}{\Delta_0^3} = \tfrac{1}{2}a'\Sigma B^{(i)}\cos i(T - P),$$

$$\frac{\alpha^2}{\Delta_0^5} = \tfrac{1}{2}a'\Sigma C^{(i)}\cos i(T - P),$$

$$\frac{\alpha^3}{\Delta_0^7} = \tfrac{1}{2}a'\Sigma D^{(i)}\cos i(T - P),$$

* *Annales de l'Obs. de Paris*, Vol. I. The terms of the eighth order have been computed by Boquet, *ib.* Vol. XIX.

and*
$$\frac{\alpha^4}{\Delta_0^9} = \tfrac{1}{2}a'\Sigma \exists^{(i)} \cos i(T-P),$$
$$\frac{\alpha^5}{\Delta_0^{11}} = \tfrac{1}{2}a'\Sigma ꟻ^{(i)} \cos i(T-P),$$

or, in general,
$$\frac{\alpha^{\frac{s-1}{2}}}{\Delta_0^s} = \tfrac{1}{2}a'\Sigma\beta_s^{(i)} \cos i(T-P).$$

If $K^{(i)}$ be any one of these coefficients, put
$$K_p^{(i)} = \frac{1}{\lfloor p}\alpha^p \frac{d^p}{d\alpha^p} K^{(i)}, \qquad \beta_{s,p}^{(i)} = \frac{1}{\lfloor p}\alpha^p \frac{d^p}{d\alpha^p}\beta_s^{(i)}.$$

Then, when the functions $\beta_{s,p}^{(i)}$ have been calculated, we have all the materials for obtaining the functions P_p.

Leverrier, besides the notation just given, uses certain functions of $A, B, \ldots$, in the development. The notation adopted here is the same as his. It is

$$(21) \qquad \left\{\begin{aligned} E^{(i)} &= \tfrac{1}{2}\,(B^{(i-1)} + B^{(i+1)}),\\ G^{(i)} &= \tfrac{3}{8}\,(C^{(i-2)} + 4C^{(i)} + C^{(i+2)}),\\ H^{(i)} &= \tfrac{5}{16}\,(D^{(i-3)} + 9D^{(i-1)} + 9D^{(i+1)} + D^{(i+3)}),\\ L^{(i)} &= \tfrac{3}{4}\,(C^{(i-2)} + C^{(i)}),\\ S^{(i)} &= \tfrac{15}{16}\,(D^{(i-3)} + 3D^{(i-1)} + D^{(i+1)}),\\ T^{(i)} &= \tfrac{15}{16}\,(D^{(i-3)} + D^{(i-1)}).\end{aligned}\right.$$

Of these, it has been rarely necessary to use any but $E^{(i)}$, $L^{(i)}$, $G^{(i)}$.

As powers of the $A^{(i)}$, $B^{(i)}$ do not occur in any of the formulae which will be used, the brackets round the i in the index will be omitted. For brevity we shall put
$$\beta_{s,0}{}^i = \beta_s^i,$$
and when the index i can be omitted in any equation without causing confusion, it will be dropped.

The coefficients in Leverrier's expansion of $1/\Delta$ are all functions of e', e'', γ'' and the A_p, B_p, ... The method outlined here for deriving all the planet functions from this one expansion does not necessarily give the shortest algebraical expressions for the coefficients, but for numerical computation, which is the principal end in view, these expressions have this advantage—that they require very little use of logarithm tables. The calculations consist mainly of additions, subtractions and multiplications by integers less than 100, and the functions are read straight from Leverrier's expansion. Moreover,

* The reversed letters $\exists$, ꟻ are used to distinguish from other functions defined on this page; Leverrier does not need the two last expansions.

it is possible to see almost immediately when the terms in a given coefficient become insensible. For example, in the case of the great inequality due to Venus, argument $l+16T-18V$, the principal term has the factor γ''^2 and argument $l+16T-18V-2h''$, and the terms of order γ''^8 are just sensible. The following table of the coefficients actually calculated shows immediately where the calculations can be stopped. The first column gives the order of the particular portion of the coefficient, and the remaining columns the parts of this order, contributed by P_1, $\frac{1}{2}(P_2+P_3)$, $\frac{1}{2}(P_2-P_3)$, respectively.

Argument $l+16T-18V-2h''$.

Order	P_1	$\frac{P_2+P_3}{2}$	$\frac{P_2-P_3}{2}$
γ''^2	$-15''\cdot89$	$-1''\cdot28$*	$+1''\cdot19$*
γ''^4	$+\ 1\cdot65$	$+\ \cdot17$	$-\ \cdot09$
γ''^6	$-\ \cdot11$	$-\ \cdot01$	$\cdot00$
γ''^8	$+\ \cdot01$		
$e'^2\gamma''^2$	$+\ \cdot27$	$\cdot00$	$-\ \cdot01$
$e''^2\gamma''^2$	$+\ \cdot12$	$+\ \cdot01$	$-\ \cdot01$
$e'^4\gamma''^2$	$-\ \cdot00$		
$e'^2e''^2\gamma''^2$	$-\ \cdot01$		
Sums	$-\ 13\cdot96$	$-\ 1\cdot11$	$+\ 1\cdot08$

All of these were fully calculated with the exception of that of order γ''^8, which a brief examination showed to be between $-0''\cdot005$ and $-0''\cdot015$; the terms of orders $e'^2\gamma''^4$, $e''^2\gamma''^4$, and a portion from the third term of R are of about the same size, but they were not calculated. I believe the result, $-13''\cdot99$, and the values for the other terms of this period, are accurate to within $0''\cdot05$; the additional computations necessary to obtain the final coefficient within $0''\cdot005$ are not very long, but in view of the uncertainty in the mass of Venus, which is doubtful within one per cent., and of the length of the period of the term, the present results are fully sufficient.

Calculation of the Functions $\beta_{s,p}^{(i)}$.

The known methods used for this purpose† have been modified in order to avoid the want of symmetry which makes them troublesome, and the considerable loss of accuracy which occurs for large values of i. The point to be considered in this connection is not the number of places of decimals but the number of significant figures, and the methods used here have been

* In the original essay these portions were given to be $-1''\cdot90$, $+1''\cdot16$, respectively, owing to a numerical error.

† See Tisserand, *Méc. Cél.*, Vol. I. Chap. 17; and Leverrier, *loc. cit.*

adopted, partly to obtain the same number of significant figures for every coefficient, partly to avoid numerous multiplications by incommensurable numbers (including the use of logarithms) and partly to obtain easy checks on the results, so that there shall be no doubt about the numerical accuracy of the tables giving the functions $\beta_{s,p}^{(i)}$. It is true that functions for values of i not afterwards needed have been found, but I believe that the gain in accuracy more than counterbalances this defect. The sieve has shown the maximum value of i which is likely to be required in the case of each planet, and the tables have been formed up to this maximum value of i. The advantage of the present methods has been obtained mainly by the use of the operator $\left(\alpha \frac{d}{d\alpha}\right)^n$ in preference to $\alpha^n \left(\frac{d}{d\alpha}\right)^n$.

We have the well known formula*

$$(22) \quad A^i = 2 \,.\, \frac{1 \,.\, 3 \,.\, 5 \ldots 2i-1}{2 \,.\, 4 \,.\, 6 \ldots 2i} \alpha^i \left[1 + \frac{1}{2} \cdot \frac{2i+1}{2i+2} \alpha^2 + \frac{1 \,.\, 3}{2 \,.\, 4} \cdot \frac{(2i+1)(2i+3)}{(2i+2)(2i+4)} \alpha^4 + \ldots\right],$$

it being remembered that for $i = 0$, the factor outside the square brackets is 2. In the case of Venus and Mars α^2 is not far from ·5 and the convergence is very slow. The plan adopted consisted in finding each coefficient to seven significant figures with seven place tables, so that the results might be trusted to six significant figures: this involved the calculation of the portion inside the bracket to seven places of decimals†.

A more rapidly convergent series can be obtained, especially for large values of i. Expand in powers of α^2,

$$\frac{1}{\sqrt{1-\alpha^2}} = b_0 + b_1 + b_2 + \ldots = 1 + \tfrac{1}{2}\alpha^2 + \tfrac{3}{8}\alpha^4 + \ldots$$

and put
$$\frac{1}{\sqrt{1-\alpha^2}} - b_0 - b_1 - \ldots - b_{i-1} = a_i, \qquad \frac{1}{\sqrt{1-\alpha^2}} = a_0,$$

$$\frac{1 \,.\, 3 \ldots (2i-1)}{2 \,.\, 4 \ldots 2i} \cdot \frac{1}{2i+2} = q_i.$$

Then an easy transformation gives

$$(23) \qquad A^i = 2\alpha^i \left[(2i+2)\, q_{i+1} a_0 - q_{i+1} a_1 - q_{i+2} a_2 - \ldots\right].$$

A table of the values of a^i was formed and also one of the logarithms of the q_i; the latter are rapidly obtained by making a table for $(2i-1)/2i$ and then one for $(2i+2)\, q_i$ and finally one for q_i. In this way any coefficient A^i can be easily and securely found.

* Tisserand, *loc. cit.*

† The phrase 'degree of accuracy' used in this section refers to the number of significant figures in the results and not to the number of places of decimals.

We have the relation*

$$A^i = \frac{2i-2}{2i-1}\epsilon A^{i-1} - \frac{2i-3}{2i-1}A^{i-2}, \qquad \epsilon = \frac{1}{\alpha} + \alpha,$$

which is ordinarily used to find A^i when A^{i-1}, A^{i-2} are known. But it is better to start from (23) with the highest value of i required and use the relation in the form

$$(24) \qquad A^i = \frac{2i+2}{2i+1}\epsilon A^{i+1} - \frac{2i+3}{2i+1}A^{i+2}.$$

The loss of accuracy is much less, and it diminishes with i; in the case of Venus the loss amounts to about one significant figure after six successive applications. For Venus, the values of A^i were found from (23) for $i = 43$, 42, 41; 30, 29, 28; 20, 19, 18; 10, 9, 8; 2, 1, 0; each triplet was tested by (24) and the intermediate values found from the same formula. An independent test is obtained by forming the successive first, second,... differences of $\log A^i$ which rapidly become constant when i is greater than about ten. This fact was used to correct the intermediate values where necessary. For Mars, α is a little smaller and the loss of accuracy less; but it was found more rapid† to calculate for $i = 30, 28, 26, \ldots 2, 0$ from (23) and to find the odd values from (24) with i an even number (since $\epsilon > 1$), testing with i odd. For the other planets it was sufficient to find the highest values from (22) and follow down with (24) to $i = 0$, which was computed from (22) as a test.

For the B^i we have (Tisserand, *loc. cit.*)

$$(25) \qquad \begin{cases} \frac{1}{2}(B^i + B^{i+1}) = \dfrac{\alpha}{2(1-\alpha)^2}(A^i - A^{i+1})(2i+1), \\ \frac{1}{2}(B^i - B^{i+1}) = \dfrac{\alpha}{2(1+\alpha)^2}(A^i + A^{i+1})(2i+1), \end{cases}$$

from which any pair of values of the B^i may be computed; there is however a loss of accuracy. But we have (Tisserand, *loc. cit.*) a formula which may be written

$$(26) \qquad B^{i-1} = 2iA^i + B^{i+1}$$

and which enables us to calculate the B^i with great rapidity, without any loss as soon as the two highest values of B^i have been found; moreover the loss in the latter disappears after the first step and the succeeding B^{i-1} have the same degree of accuracy as the corresponding A^i. For security, the B^i were calculated from (25) after each ten steps, and several were tested by the formula (Tisserand, *loc. cit.*)

$$B^i = \frac{2i-2}{2i-3}\epsilon B^{i-1} - \frac{2i-1}{2i-3}B^{i-2}.$$

* See Tisserand, *Méc. Cél.*, Vol. I. Chap. 17. Tisserand gives the formulae for A^i, B^i/a, C^i/a^2, ...

† The calculations for Venus had been completed before Mars was undertaken.

The greatest advantage of (26) however is the rapidity with which the calculations are performed by means of it, for it does not require the use of logarithms.

Special values of the remaining functions C^i, D^i,... were obtained from the formula (to be proved later on)

$$(27)\quad \beta_{s+2}{}^{i} = \frac{2s+1}{s+1}\cdot\frac{\epsilon}{\epsilon'^2}\beta_{s+1}{}^{i} + \frac{1}{\epsilon'^2}\frac{(i^2-s^2)}{s(s+1)}\beta_s{}^{i}, \quad \epsilon' = \frac{1}{\alpha}-\alpha, \quad \epsilon = \frac{1}{\alpha}+\alpha,$$

which involves no loss of accuracy, combined with

$$(28)\qquad \beta_{s+1}^{i-1} = \frac{1}{s}\, i\beta_s{}^{i} + \beta_{s+1}^{i+1},$$

for the intermediate values.

(It is interesting to notice that (27) gives

$$\beta_{\frac{3}{2}} = -\frac{4i^2-1}{\epsilon'^2}\beta_{-\frac{1}{2}},$$

but no use was found for this formula.)

The $\beta_{s,p}{}^{i}$ have now to be obtained. There are two problems to be considered: the first is where a *table* is to be made for all values of i up to some definite place (30 in the case of Venus* and Mars, 8 for Mercury and 6 for Jupiter). We require here a formula for special values of i, and when the coefficients for the two highest values have been found, the remainder are obtained from the obvious generalization of (28):

$$(29)\qquad \beta_{s+1,p}^{i-1} = \frac{i}{s}\beta_{s,p}{}^{i} + \beta_{s+1,p}^{i+1}.$$

This method does not give the values for $s=\frac{1}{2}$; the formula for special values of i is required throughout. The second problem arises when so few values of i are needed that it is not worth while to make a full table. Hence we require to develope formulae from which the $\beta_{s,p}{}^{i}$ can be obtained for special values of i, it being understood that the *tables* are always formed with the assistance of (29).

In the following, the index is generally omitted, since it is the same in all the formulae now to be obtained.

Remembering that

$$\Delta_0{}^{-s} = \{1-2\alpha\cos(T-P)+\alpha^2\}^{-\frac{s}{2}} = \tfrac{1}{2}\alpha'\alpha^{-(s-\frac{1}{2})}\Sigma_i\beta_s{}^{i}\cos i(T-P),$$

$$D = \alpha\frac{d}{d\alpha},\quad \epsilon' = \frac{1}{\alpha}-\alpha,\quad \beta_s{}^{-i} = \beta_s{}^{i},$$

we easily obtain, if for a moment we put $\beta_s\alpha^{-s+\frac{1}{2}} = b_s$,

$$Db_s = s(1-\alpha^2)\,b_{s+1} - sb_s,$$

$$D^2b_s = 4s^2\alpha^2b_{s+1} + i^2b_s,$$

$$(D+2s)^2b_s = 4s^2b_{s+1} + i^2b_s;$$

* Except for A, B, which were required to $i=43$.

whence

$$(30)\quad \begin{cases} D\beta_s = s(1-\alpha^2)\beta_{s+1} - \tfrac{1}{2}\beta_s, \\ (D - s + \tfrac{1}{2})^2\beta_s = 4s^2\alpha\beta_{s+1} + i^2\beta_s, \\ (D + s + \tfrac{1}{2})^2\beta_s = \dfrac{4s^2}{\alpha}\beta_{s+1} + i^2\beta_s, \end{cases}$$

giving

$$(31)\quad D^2\beta_s = 4s^2\alpha\beta_{s+1} + 2(s-\tfrac{1}{2})D\beta_s + \{i^2 - (s-\tfrac{1}{2})^2\}\beta_s,$$

$$(32)\quad (D+1)^2\beta_s = 2s(s\epsilon + \tfrac{1}{2}\epsilon')\beta_{s+1} + (i^2 - s^2 + \tfrac{1}{4})\beta_s.$$

By definition,

$$\beta_{s,p} = \frac{\alpha^p}{\lfloor p} \cdot \left(\frac{d}{d\alpha}\right)^p \beta_s$$
$$= \frac{1}{\lfloor p} D(D-1)(D-2)\dots(D-p+1)\beta_s,$$

whence, if k be any number independent of α,

$$(33)\quad (D+k)\beta_{s,p} = (p+1)\beta_{s,p+1} + (p+k)\beta_{s,p},$$

$$(34)\quad (D+k)^2\beta_{s,p} = (p+1)(p+2)\beta_{s,p+2} + (p+1)(2p+2k+1)\beta_{s,p+1} + (p+k)^2\beta_{s,p}.$$

Again, from (30),

$$(35)\quad (D - s + \tfrac{1}{2})^2\beta_{s,p} = \frac{1}{\lfloor p} D(D-1)\dots(D-p+1)(4s^2\alpha\beta_{s+1} + i^2\beta_s)$$
$$= 4s^2\alpha(\beta_{s+1,p} + \beta_{s+1,p-1}) + i^2\beta_{s,p}.$$

Combining this with (34), after putting $k = -s + \tfrac{1}{2}$, we find

$$(36)\quad p(p+1)\beta_{s,p+1} = 4s^2\alpha(\beta_{s+1,p-1} + \beta_{s+1,p-2}) - p(2p-2s)\beta_{s,p} + \{i^2 - (p-s-\tfrac{1}{2})^2\}\beta_{s,p-1}.$$

The material contained in these equations is sufficient to find all the $\beta_{s,p}$ rapidly, when the β_s are known.

We have, in fact,

$$A_1 = \tfrac{1}{2}\epsilon' B_0 - \tfrac{1}{2}A_0,$$
$$B_1 = \tfrac{3}{2}\epsilon' C_0 - \tfrac{1}{2}B_0,$$
$$C_1 = \tfrac{5}{2}\epsilon' D_0 - \tfrac{1}{2}C_0,$$

....

$$2A_2 = \alpha B_0 - A_1 + i^2 A_0,$$
$$2B_2 = 9\alpha C_0 + B_1 + (i^2-1)B_0,$$
$$2C_2 = 25\alpha D_0 + 3C_1 + (i^2-4)C_0,$$

....

$$6A_3 = \alpha(B_1 + B_0) - 6A_2 + (i^2-1)A_1,$$
$$6B_3 = 9\alpha(C_1 + C_0) - 2B_2 + i^2 B_1,$$
$$6C_3 = 25\alpha(D_1 + D_0) + 2C_2 + (i^2-1)C_1,$$

....

$$12A_4 = \alpha(B_2 + B_1) - 15A_3 + (i^2-4)A_2,$$
$$12B_4 = 9\alpha(C_2 + C_1) - 9B_3 + (i^2-1)B_2,$$

....

The required functions $\beta_{s,p}\,(p>0)$ are obtained from these without loss of accuracy.

The functions $\{(D+1)^2-i^2\}\,\beta_{s,p}$, $(D^2-i^2)\,\beta_{s,p}$ are required to find the $P_1, P_6, P_7, \ldots$ (page 24). If these were obtained from (34) with the values of $\beta_{s,p}$ just found they would in general entail a loss of accuracy of the order of i units in the last calculated figure.

Equations (31), (32) enable us to calculate them without loss when $p=0$. When $p>0$, we have from (34), (36),

$$\{(D+1)^2-i^2\}\,\beta_{s,p}=4s^2\alpha\,(\beta_{s+1,p}+\beta_{s+1,p-1})+(p+1)(2s+1)\,\beta_{s,p+1}+(s+\tfrac{1}{2})(2p-s+\tfrac{3}{2})\,\beta_{s,p}.$$

In particular,

$$\{(D+1)^2-i^2\}\,A_0=\frac{B_0}{\alpha},$$

$$\{(D+1)^2-i^2\}\,B_0=\left(\frac{6}{\alpha}+3\alpha\right)C_0-2B_0,$$

$$\{(D+1)^2-i^2\}\,C_0=\left(\frac{15}{\alpha}+10\alpha\right)D_0-6C_0,$$

....

$$\{(D+1)^2-i^2\}\,A_p=\alpha\,(B_p+B_{p-1})+2\,(p+1)\,A_{p+1}+(2p+1)\,A_p,$$
$$\{(D+1)^2-i^2\}\,B_p=9\alpha\,(C_p+C_{p-1})+4\,(p+1)\,B_{p+1}+2\,.\,2pB_p,$$
$$\{(D+1)^2-i^2\}\,C_p=25\alpha\,(D_p+D_{p-1})+6\,(p+1)\,C_{p+1}+3\,(2p-1)\,C_p,$$

which are rapidly obtained, since the multiplication by α will have been made in finding the $\beta_{s,p}$.

In a similar way we obtain from (31), (34), (36) for outer planets,

$$(D^2-i^2)\,A_0=\alpha B_0,$$
$$(D^2-i^2)\,B_0=9\alpha C_0+2B_1-B_0,$$
$$(D^2-i^2)\,C_0=25\alpha D_0+4C_1-4C_0,$$

....

$$(D^2-i^2)\,A_p=\alpha\,(B_p+B_{p-1}),$$
$$(D^2-i^2)\,B_p=9\alpha\,(C_p+C_{p-1})+2\,(p+1)\,B_{p+1}+(2p-1)\,B_p,$$
$$(D^2-i^2)\,C_p=25\alpha\,(D_p+D_{p-1})+4\,(p+1)\,C_{p+1}+2\,(2p-2)\,C_p.$$

....

It has been found convenient to alter certain of the formulae (21) a little. We have

$$iA^i=\tfrac{1}{2}B^{i-1}-\tfrac{1}{2}B^{i+1},$$
$$iB^i=\tfrac{3}{2}C^{i-1}-\tfrac{3}{2}C^{i+1}.$$

Applying the first of these to the definition of E^i, the second to that for L^i, and both to that for G^i, it is easy to show that

$$(37)\qquad \left\{\begin{aligned} &E^i=B^{i-1}-iA^i=E^{-i},\\ &L^i=\tfrac{3}{2}C^i+\tfrac{1}{2}(i-1)\,B^{i-1},\\ &G^i=\tfrac{9}{4}C^i-\tfrac{1}{2}B^{i-1}+\frac{i\,(i+1)}{2}A^i.\end{aligned}\right.$$

These, equally with (21), can be operated on by any function of D and consequently they are true when the suffix p has been attached to the symbols.

Solution of a Difficulty.

From the series for A_p, or otherwise, it is obvious that $(D-i)^q A^i$ is of the same order of magnitude in general as A^i. Hence when i is large and this function is calculated from the tables it appears as the difference of two nearly equal numbers, and a considerable loss of accuracy results. The same fact is true of all the functions $A_p, B_p, \ldots$, but for a given value of i the loss of accuracy is smaller as p increases and as we go along the series $A, B, C, \ldots$. Although the number of places used in the tables has been found to be sufficient for the purposes of this essay, I shall give a method showing how the difficulty may be overcome.

We have (Tisserand, *loc. cit.*):

$$A^i = \frac{2}{\pi} \alpha^i \int_0^\pi \frac{\sin^{2i}\psi}{(1-\alpha^2\sin^2\psi)^{\frac{1}{2}}} d\psi.$$

Hence
$$(D-i)\,A^i = \frac{2}{\pi} \alpha^{i+2} \int_0^\pi \frac{\sin^{2i+2}\psi}{(1-\alpha^2\sin^2\psi)^{\frac{3}{2}}} d\psi,$$

$$(D-i-1)\,A^{i+1} = \frac{2}{\pi} \alpha^{i+3} \int_0^\pi \frac{\sin^{2i+4}\psi}{(1-\alpha^2\sin^2\psi)^{\frac{3}{2}}} d\psi.$$

Therefore

$$\begin{aligned}(D-i)\,A^i &= \alpha A^{i+1} + \alpha(D-i-1)\,A^{i+1} \\ &= \alpha A^{i+1} + \alpha^2 A^{i+2} + \ldots + \alpha^j A^{i+j} + \alpha^j (D-i-j)\,A^{i+j}.\end{aligned}$$

The values of A^i, $DA^i = A_1{}^i$ have been calculated to seven significant figures, and the ratio of $(D-i-1)\,A^{i+1}$ to $(D-i)\,A^i$ approaches the limit α as i increases. Hence if p is the number of units loss of accuracy in the last significant figure of $(D-i-j)\,A^{i+j}$, the number of units loss in $(D-i)\,A^i$ is $p\alpha^{2j}$ approximately, which can always be reduced to unity or to a gain by taking j large enough. For values of i less than i' we must take $\alpha^{2j} < i'$ if we are to lose no accuracy in $(D-i)\,A^i$. In the case of Venus, with $i'=30$, this gives $j=5$, $i=25$.

It is obvious that this principle can be extended to any of the functions previously used by applying the preceding formulae, but since the practical applications did not demand it I shall not develope the formulae here.

It does not seem possible to obtain a general formula for $(D-i)\,\beta_{s,p}{}^i$ in terms of $\beta_{s,p}{}^i$ without loss of accuracy otherwise than by the principle of the method outlined here; i.e., by referring forward to some higher value of i.

The corresponding difficulty in Radau's work appears in the form

$$A^i - \alpha A^{i-1},\ A^i - 2\alpha A^{i-1} + \alpha^2 A^{i-2}, \ldots.$$

Calculation of the Coefficients M_i and of their Derivatives with respect to n.

The most arduous part of the work has consisted in the computation of the derivatives of r^2, $x^2 - y^2 + 2ixy = u_0^2$ with respect to n, to the degree of accuracy which was demanded. The only complete literal expansion in powers of m is that given by Delaunay, and his results suffer from two serious defects for this purpose, the one in the fact that the coefficients for the parallax on which the desired quantities chiefly depend are not taken far enough, and the other in the doubts as to the degree of convergence of the series along powers of m. Both may to a certain extent be remedied by a judicious use of the numerical results given in my Lunar Theory. So far as I know, the only method for calculating these derivatives from numerical results in which the value of m has been substituted, without having recourse to algebraical series in powers of m, is one which I gave in 1903*; this method, although tedious and difficult for numerical application, has perforce been adopted in the absence of any other plan.

The method essentially consists in finding three independent relations between the coordinates and their derivatives with respect to the six elements, expressed in the form of algebraic or differential equations. From these three relations it is possible to obtain the derivatives of the coordinates with respect to one of the elements in terms of the others. No approximation processes are necessary—only multiplications of series and a quadrature for each coordinate. Further, the results naturally appear in the form in which they are required in the disturbing function, namely, as derivatives of the rectangular coordinates, and the value of $\frac{dc_1}{dn}$, on which the principal term in the equations of variations mainly depends, also arises naturally. The order with respect to e, k to which the results can be calculated is one less than that to which the coordinates have been obtained, but it was not found necessary to proceed to high orders.

The formulae are constructed to solve the problem: Given the lunar coordinates and their derivatives with respect to c_2, c_3, w_1, w_2, w_3, to find the derivatives with respect to n, it being supposed that the coordinates are expressed in terms of these six elements. My lunar theory enables one to get all the data of the problem accurately. It is expressed in terms of the constants e, k, which are left arbitrary; but the second and third of equations (8) together with the values of c_2, c_3 in terms of these constants enable one easily to find the derivatives with respect to c_2, c_3 from those with respect to e, k, or *vice versâ*. The method given for performing the calculations was adopted almost without change. The results will be found in the

* *Trans. Amer. Math. Soc.* Vol. IV. pp. 234—248.

collection of tables. The details by which the functions chiefly needed, namely,

$$\frac{d}{dn}(r^2 - 3z^2), \quad \frac{d}{dn}(\mathrm{u}_0^2), \quad \frac{d}{dn}(\mathrm{u}_0 z),$$

were found directly instead of the functions for which the formulae were given, namely,

$$\frac{d}{dn}(\mathrm{u}_0), \ \frac{d}{dn}(z),$$

will be omitted*.

The greater part of the work consisted in multiplications of series of the same nature as those necessary in my lunar theory. The calculation of

$$r^2 - 3z^2, \ \mathrm{u}_0^2, \ \mathrm{u}_0 z$$

consists entirely of such multiplications. The multiplications by

$$\frac{a'^2}{r'^2}, \ \frac{a'^2}{r'^2} e^{\mp 2j(V' - T)}, \ \ldots,$$

are quickly performed. Certain coefficients of orders higher than those given in the tables were required. These were obtained from the complete results already quoted for u_0, z: their derivatives with respect to n were multiplied by such small coefficients that a calculation of them was obviously unnecessary. Thus in all this work the effect of the solar perturbations in the moon's motion on the coefficients of the planetary terms has been fully taken into account. It may be stated here that the derivatives with respect to n are only required either in the terms of comparatively short period (say, a few years or less), or in those terms which do not contain the argument w_1. A glance at the final form of the equations of variations will make this statement evident.

* Some errors in the formulae of the paper referred to were found. The integrated equation (*C*) p. 246, requires an added constant, contrary to the statement there made: this constant may be determined by the equation numbered (14) in the paper; and the sign before the last term of (*E*) p. 247 should be changed, as well as the first negative sign in equation (15), p. 244.

SECTION IV.

A SIEVE FOR THE REJECTION OF INSENSIBLE COEFFICIENTS.

THE periodic inequalities due to the direct action of the planets are usually divided into two classes, which have received the names 'long-period' and 'short-period.' These names are somewhat misleading in this connection, because there is no sharp line of division. I shall divide them into *primary* and *secondary* inequalities: the former are those which arise from the substitution of $w_1 + \delta w_1$ for the *non-periodic term* of V; the secondary inequalities will be defined as those which arise from the substitution of the variable values of the elements in the *periodic terms*.

The majority of the primary inequalities are of long period—a year or more; but there are classes of them having sensible coefficients with periods of a month or less. Nearly all the secondary inequalities are of short period. Thus the primary inequalities in the longitude are obtained from

$$\delta V = \delta w_1,$$

and the secondary inequalities from

$$\delta V = \left(\frac{dV}{dw_1} - 1\right) \delta w_1 + \frac{dV}{dw_2} \delta w_2 + \frac{dV}{dw_3} \delta w_3 + \frac{dV}{dn} \delta n + \frac{dV}{dc_2} \delta c_2 + \frac{dV}{dc_3} \delta c_3.$$

The coefficients of the secondary inequalities will be smaller than the coefficient of the corresponding primary inequality, in general; an examination of the final form of the equations of variations and of the values of the coefficients in the derivatives of V shows this immediately. An exception will occasionally occur in the terms

$$\frac{dV}{dc_2} \delta c_2, \quad \frac{dV}{dw_2} \delta w_2.$$

The largest term in $\frac{dV}{dc_2}$ (see page 12) has a coefficient 40; the coefficient in δc_2 is

$$7 \times 10^{-6} \cdot \frac{s}{\lambda_1} i_2 C,$$

that in δw_1 is

$$c\left(1+\frac{f's}{f\lambda_1}\frac{A_1}{A}\right).$$

Hence the ratio of the largest coefficient in $\frac{dV}{dc_2}\,\delta c_2$ to that of δw_1 is

$$\rho=\frac{1}{2}\frac{28}{10^5}\cdot\frac{si_2}{\lambda_1}\cdot\frac{1}{1+\frac{f's}{f\lambda_1}\cdot\frac{A_1}{A}}=\frac{14i_2}{10^5\frac{\lambda_1}{s}+2\frac{A_1}{A}},$$

since

$$\frac{f'}{f}=\frac{2}{10^5}.$$

As $\frac{A_1}{A}$ is never very small (it is in most cases a little greater than unity) the value of ρ will be large only if the two terms of the denominator are approximately equal in magnitude and opposite in sign. The sieve which follows does not contemplate this 'accidental' approximate vanishing of the denominator, so that it would not reject the inequality in δw_1 for this cause. Hence the maximum value of ρ would appear from the sieve as the greater of the two values

$$7i_2\frac{A}{A_1}\quad\text{and}\quad\frac{14i_2 s}{10^5\lambda_1}<\frac{s}{100},$$

since $\lambda_1=i_1+{\cdot}01486i_2-{\cdot}003744i_3$ and its least value when $i_2\neq 0$ is taken to be ·01486 in the sieve. The terms in which the former is the greater are all calculated without reference to the sieve. Those in which the latter is the greater demand that $s>100''$ for the secondary inequality to be rejected by the sieve for δw_1 when it ought to be retained. The sieve shows that such inequalities can only occur with the lunar arguments $2\mathrm{D}-2l$, $\mathrm{D}-l$, and these have been separately examined. The term $\frac{dV}{dw_2}\,\delta w_2$ is of the same order approximately, and the same argument applies.

A similar argument applied to the terms containing $2\mathrm{D}-2\mathrm{F}$, $\mathrm{D}-\mathrm{F}$ shows that the sieve would not reject the secondary inequalities in longitude as long as $s<2000''$, and as the upper limit for the sieve is $s=3500''$, a secondary inequality with a maximum coefficient of $0''{\cdot}02$ might have been rejected by it. But the inequalities for $s>2000''$ were separately examined.

The secondary inequalities in the latitude of terms containing $2\mathrm{D}-2\mathrm{F}$, $\mathrm{D}-\mathrm{F}$ are more important than those in the longitude. A similar argument applied to them shows that the primary might be neglected by the sieve if $s>60''$ when the secondary should be retained. Consequently a separate examination of these terms was also undertaken.

We have then to examine the primary inequalities. Those which have periods of less than a year do not need a sieve. The terms in the moon's coordinates which, in combination with the planet's coordinates can give sensible terms, are few in number and they are taken one by one according to the sizes of their coefficients with all possible terms arising from the planet's coordinates until they become insensible; the number of the moon terms thus required is quite small, as the results show, and any omissions would arise only from numerical errors. The sieve which follows is therefore constructed for periods greater than a year. Some limitation had to be placed on the maximum period to be included; it was taken at 3500 years, corresponding to a value of s equal to $1''$. But values of s less than $1''$ which arose from the equations constituting the sieve were also included.

The results show that though there will be inequalities of periods greater than 3500 years, their order is so high that it is extremely doubtful if their coefficients would be sensible. In any case the period is so long that they would scarcely be observable within historic times. It is to be noted that for very long periods it is better to expand such terms in powers of t when we wish to examine their effect on the observed position of the moon. Thus:

$$C \sin(qt + q') = \alpha_0 + \alpha_1 t + \alpha_2 t^2 + \ldots.$$

The first two terms would then constitute an addition to the expression for the moon's mean longitude, which, being one of the observed quantities, would not be affected: the values of α_0, α_1 could hardly be large enough for this change in the meaning of w_1 to affect the periodic terms. The succeeding terms would constitute an addition to the secular acceleration, but it is quite improbable that there is any coefficient great enough to affect it by so much as $0''\cdot 1$ per century.

It has not hitherto been found possible to construct any kind of analysis which will give a criterion for the magnitude of the coefficients of the planetary terms without having recourse early to numerical developments. There are many thousands of inequalities whose periods suggest that their coefficients might be sensible but which on computation are found to be exceedingly small. Even the roughest approximation would be a laborious process, and it seemed advisable at the outset to construct formulae which could be rapidly applied. Fortunately, for the two most troublesome planets —Venus and Mars—numerical developments were available for this purpose. Newcomb has given* the expansions of $\frac{1}{\Delta^3}$, $\frac{1}{\Delta^5}$ in cosines and sines of $iT - jP$ up to $i = 29$, with a sufficient number of values of $i - j$†. The numerical values of all quantities have been substituted at the outset, so that

* *Amer. Eph. Papers*, Vol. v. pt. 3.

† The order of any term with reference to the eccentricities of the earth and planet and the inclination of the planet is $|i - j|$.

there is no possibility of using the method of Section II. But as the arguments are developed in multiples of the mean longitudes it is still possible to differentiate or integrate with respect to these, on the understanding that we consider the arguments as functions of the mean longitudes and of the longitudes of the perigees and of the node (not of the mean longitudes and mean anomalies as in Section II.). For Mars the development had not been carried beyond $i = 11$ for $\frac{1}{\Delta^3}$, and not completely to that point; but a process of extrapolation and a knowledge of the general law of decrease enabled one to extend the table for the purposes of a sieve sufficiently far: a liberal margin was left for errors in this extrapolation. For the other planets, the general law of decrease, which approximately holds when i is sufficiently great, namely $K^i/K^{i+1} = \alpha$, is so soon reached owing to the smallness of α, that there was no difficulty in forming a rough table.

There are two problems to be dealt with: first to form a table from which all possible periods can be rapidly obtained; second, to obtain formulae for finding easily the order of magnitude of the coefficient which will result in the moon's longitude. The process which follows is divided into several parts:—

(*a*) A formula for the coefficient in longitude when we know N (the coefficient from Newcomb's tables), s (the number of seconds in the daily motion of its argument), k (the order of the term with respect to the lunar eccentricity and inclination). Modification for the planets other than Venus.

(*b*) A formula for the order with respect to e, γ associated with a given multiple of w_1, w_2, w_3 in order to obtain k.

(*c*) A method of constructing the period corresponding to each multiple of l, g, h to be considered, to obtain s.

(*d*) The terms which were not rejected. These will be found with the numerical results.

The only part of the actual computations which required special care was that giving the periods. It was found, however, throughout the numerical work that, the plan once formed and in working order, the size of the coefficients could be almost guessed at a glance, so that the computations were themselves of the nature of a test of accuracy, and errors tended rather to make the coefficient too large than too small. The test was gone through twice. The details which resulted in the table of coefficients retained by the sieve for accurate computation will be omitted.

Construction of the Sieve.

We have, as in Section II,

$$\xi = -r' + (1-\gamma''^2)\,r''\cos(V''-V') + \gamma''^2 r''\cos(V''+V'-2h''),$$

$$\eta = (1-\gamma''^2)\,r''\sin(V''-V') - \gamma''^2 r''\sin(V''+V'-2h''),$$

$$\zeta = 2\gamma''\sqrt{1-\gamma''^2}\,r''\sin(V''-h''),$$

$$\Delta^2 = r''^2 - r'^2 - 2r'\xi.$$

Hence, putting $E = \dfrac{dr'}{dT}, \quad N = r'\dfrac{dV'}{dT},$

we obtain $\Delta\dfrac{d\Delta}{dT} = -\xi E - \eta N,$

$$(38)\qquad \frac{1}{s}\frac{d}{dT}\left(\frac{1}{\Delta^s}\right) = \frac{\eta N + \xi E}{\Delta^{s+2}}.$$

Multiplying the latter equation by $\xi E - \eta N$, $\xi N + \eta E$ successively, and making use of the equation $\Delta^2 = \xi^2 + \eta^2 + \zeta^2$ in the form*

$$-\frac{\xi^2}{\Delta^{s+2}} = -\left(\frac{1}{\Delta^s} - \frac{\zeta^2}{\Delta^{s+2}}\right) + \frac{\eta^2}{\Delta^{s+2}},$$

we find, after a few simple operations,

$$(39)\qquad \left\{\begin{aligned} \frac{\xi^2-\eta^2}{\Delta^{s+2}} &= \frac{N^2-E^2}{N^2+E^2}\left(\frac{1}{\Delta^s} - \frac{\zeta^2}{\Delta^{s+2}}\right) - \frac{2(\eta N - \xi E)}{s(N^2+E^2)}\frac{d}{dT}\left(\frac{1}{\Delta^s}\right) \\ \frac{2\xi\eta}{\Delta^{s+2}} &= -\frac{2NE}{N^2+E^2}\left(\frac{1}{\Delta^s} - \frac{\zeta^2}{\Delta^{s+2}}\right) + \frac{2(\xi N + \eta E)}{s(N^2+E^2)}\frac{d}{dT}\left(\frac{1}{\Delta^s}\right). \end{aligned}\right.$$

Now the principal terms in the expansion of N, E in powers of e' are a', $a'e'$, and the equation (38) shows that when the multiple of T considered in the expansion of $\dfrac{1}{\Delta^s}$ in cosines is not very small, $\dfrac{1}{\Delta^s}$ is small compared with $\dfrac{1}{\Delta^{s+2}}$ (s positive). Hence we shall evidently be taking the worst case if we consider

$$\frac{\xi^2-\eta^2}{\Delta^5},\quad \frac{2\xi\eta}{\Delta^5} \text{ as of the orders } \frac{4\eta}{3}\frac{d}{dT}\left(\frac{1}{\Delta^3}\right),\quad \frac{4\xi}{3}\frac{d}{dT}\left(\frac{1}{\Delta^3}\right).$$

Further,

$$\xi\frac{d}{dT}\left(\frac{1}{\Delta^3}\right) = \frac{d}{dT}\left(\frac{\xi}{\Delta^3}\right) - \frac{1}{\Delta^3}\frac{d\xi}{dT}$$

$$= \text{order of } \frac{d}{dT}\left[\frac{r'^2 - r''^2}{2r'\Delta^3} - \frac{1}{\Delta}\right],$$

* The symbol s is only used in this connection on this page, and it will not be confused with the s in the equations of variations.

which, for the worst case—that of Venus where $a''^2 = 2a'^2$ approximately—gives

$$\xi \frac{d}{dT}\left(\frac{1}{\Delta^3}\right) = \text{order of } \frac{a'}{4}\frac{d}{dT}\left(\frac{1}{\Delta^3}\right).$$

The order of $\eta \frac{d}{dT}\left(\frac{1}{\Delta^3}\right)$ is approximately the same.

Consider next the function

$$\frac{1}{\Delta^3} - \frac{3\zeta^2}{\Delta^5}.$$

Here ζ is of the order $2\gamma''$. Suppose that these functions be expanded in powers of γ''^2 and then in cosines. For Venus $\gamma''^2 < \frac{1}{1000}$ and therefore for a given multiple p of T in which $p < p_1$, the terms factored by γ''^2 will be less than those independent of γ''^2: the value of p_1 which would make them greater, is seen in the course of working the method to be too large to need consideration. Hence for coefficients which do not contain γ''^2 as a factor*, it is sufficient to take $\frac{1}{\Delta^3}$: for those which contain γ''^2 as a factor we should take $\frac{6\gamma''^2}{\Delta^5}$, but this being less than $\frac{6}{\Delta^3}$ within the limits, the latter was used.

The first term of the disturbing function R in Section II., when written in terms of the real variables, is

$$R = m''\left\{\tfrac{1}{4}(r^2 - 3z^2)\left(\frac{1}{\Delta^3} - \frac{3\zeta^2}{\Delta^5}\right) + \tfrac{3}{4}(x^2 - y^2)\left(\frac{\xi^2 - \eta^2}{\Delta^5}\right) + 3xy\frac{\xi\eta}{\Delta^5} + 3z\zeta\frac{x\xi + y\eta}{\Delta^5}\right\},$$

in which the first factor is the part depending on the moon. If we take $e = \frac{1}{10}$, $\gamma = \frac{1}{10}$ instead of their actual values $\frac{1}{18}$, $\frac{1}{22}$, and in general take no account of the fact that $m = \frac{1}{13}$, we shall not be making the coefficients too small if we consider $r^2 - 3z^2$, $x^2 - y^2$, $2xy$, $2zx$, $2zy$ as of the order $a^2 10^{-k}$ where k is the order of the term considered with reference to e, γ. Remembering that a combination of two sines or cosines introduces the factor $\frac{1}{2}$, we find the following results, in which the factors have been increased slightly to make them suitable whole numbers.

First term of R, order $\frac{1}{5}\frac{m''a^2}{\Delta^3} 10^{-k}$ when γ''^2 is not a factor,

$\frac{m''a^2}{\Delta^3} 10^{-k}$ when γ''^2 is a factor.

Second and third terms of R, order $\frac{1}{5}\frac{d}{dT}\frac{m''a^2}{\Delta^3} 10^{-k}$,

fourth term of R, order $\frac{1}{10}\frac{m''a^2a'^2}{\Delta^5} 10^{-k}$.

* The parts depending on γ''^2 always diminish the coefficient.

Next, the primary inequalities are found with

$$\delta w_1 = \left[\lambda_1 \frac{fA}{s^2} + \frac{f'}{s} \frac{n}{\text{a}^2} \frac{d}{dn} (A\text{a}^2) \right] \sin (qt + q'),$$

where

$$R = \frac{1}{4} \frac{m''}{m'} n'^2 \text{a}^2 A \cos (qt + q') = \frac{1}{4} \frac{m''}{m'} n'^2 \text{a}^2 A \cos (i_1 w_1 + i_2 w_2 + i_3 w_3 + q''t + q'''),$$

$$\lambda_1 = - i_1 + \cdot 01486 i_2 - \cdot 003744 i_3,$$

$$f = [12\cdot 30] \frac{m''}{m'}, \quad f' = [7\cdot 62] \frac{m''}{m'}.$$

There are several cases to be considered.

Case 1. $i_1 = i_2 = i_3 = 0$. The second term of δw_1 with the first term of R since the other terms of R will have the factor m^2 at least;

$$\left| \frac{n}{\text{a}^2} \cdot \frac{d}{dn} (A\text{a}^2) \right| \doteq \tfrac{4}{3} \, | A |, \quad k = 0.$$

Coefficient in δw_1 is of order

$$\frac{10^{7\cdot 2}}{s} \frac{m''}{m'} \left(\text{coef. in } \frac{a'^3}{\Delta^3} \right), \quad \gamma''^2 \text{ not a factor,}$$

$$\frac{10^{7\cdot 9}}{s} \frac{m''}{m'} \left(\text{coef. in } \frac{a'^3}{\Delta^3} \right), \quad \gamma''^2 \text{ a factor.}$$

Case 2. $i_1 = 0$, i_3 even. The first term of δw_1 combined with the second or third terms of R;

Coefficient in δw_1 is of order

$$\frac{10^{9\cdot 8 - k}}{s^2} \frac{m''}{m'} \left(\text{coef. in } \frac{d}{dT} \frac{a'^3}{\Delta^3} \right).$$

Case 3. $i_2 = - i_1$, $i_1 \gtreqless 3$, i_3 even. The first term of δw_1 combined with the first term of R;

Coefficient in δw_1 is of order

$$\frac{10^{12\cdot 8 - k}}{s^2} \frac{m''}{m'} \left(\text{coef. in } \frac{a'^3}{\Delta^3} \right).$$

The other cases, depending on the way in which w_1, w_2, w_3 enter into the argument, are treated in like manner.

I give next the final form of the equations used in the case of Venus. Here $\frac{m''}{m'} = \frac{1}{408,000}$. Newcomb tabulates $24 \frac{a'^3}{\Delta^3}$, $24 \frac{a'^5}{\Delta^5}$. If N_3, N_5 are the coefficients derived from Newcomb's tables for these functions, p the multiple of T present, C the coefficient in δw_1 expressed in seconds of arc, we find the following expressions for $\log_{10} C$; the arguments are put into Delaunay's notation $(il + i'g + i''h)$. Additions have been made to the numbers for simplicity in the results and for parts arising from the terms in the moon coordinates containing e'.

Argument		*Log C*
l, g, h absent		$1{\cdot}5 - \log s + \log N_3$,
l absent, $i' + i''$, even	greater of	$\begin{cases} 4{\cdot}5 - k - 2\log s + \log p + \log N_3, \\ 3{\cdot}5 - k - 2\log s + \log N_3, \end{cases}$
„ $i' + i''$, odd		$3{\cdot}5 - k - 2\log s + \log N_5$,
l present, $i' = i'' = 0, 1$		$6{\cdot}0 - k - 2\log s + \log N_3$,
„ i'' even		$5{\cdot}0 - k - 2\log s + \log p + \log N_3$,
„ i'' odd		$5{\cdot}0 - k - 2\log s + \log N_5$.

From the second term of R, which produces terms containing the factor $\frac{a}{a'}$, by a similar procedure I find

Argument	*Log C*
$\lvert i \rvert = 1$, $i' + i''$ even	$2{\cdot}3 - k - 2\log s + \log N_5$,
$\lvert i \rvert = 3$, „	$2{\cdot}0 - k - 2\log s + 2\log p + \log N_3$,
„ $i' + i''$ odd	$1{\cdot}5 - k - 2\log s + \log p + \log N_5$.

For the other planets, similar expressions can be obtained. Except in the case of Mars as far as $i = 11$, no tables giving N_3, N_5 are available. It is therefore necessary to construct a method for the coefficients in $\frac{a'^3}{\Delta^3}$, $\frac{a'^5}{\Delta^5}$. As stated above, the parts of these functions independent of the eccentricities of the earth and planet can be expanded rapidly for a rough approximation. The terms which contain the eccentricities as factors are in general of the orders $(pe')^q$ or $(pe'')^q$ (where p is the multiple of T present, q the order with respect to e', e'') compared with the order of the term with argument $p(T - P)$, as a reference to Leverrier's development shows. A margin was, however, left for the occurrence of numerical factors independent of $\frac{a''}{a'}$, p, but depending to a small extent on q. Jupiter was the only planet in which the approximations to the coefficients came close to their correct values; and this occurred where the approximation was comparatively simple, namely, for small values of p: a smaller margin has been necessary.

The coefficients which the sieve gave as greater than $0''{\cdot}01$ will be found in the collection of numerical results.

With a knowledge of the greatest values of N_3, N_5 (which arise with the lowest values of p) it is a simple matter to find the maximum value of s to be considered in any given case. In fact, the maxima of s are found by equating the above expressions to -2, since the least value of C to be considered has been taken at $0''{\cdot}01$. Many limitations which it is not necessary to specify in detail appeared during the progress of the work; e.g., if the multiple of l present in the part of R independent of $\frac{a}{a'}$ is 2, it is only necessary to consider values of p lying between 25 and 50, and then only those in which the sum of the multiples of P, $-T$ is less than 5*.

* These statements are always made with coordinates referred to axes following the mean or true place of the sun.

Lowest orders with reference to e, k associated with the Lunar Arguments.

If any argument contains w_1, w_2, w_3 in the form*

$$i_1 w_1 + i_2 w_2 + i_3 w_3 = il + i'g + i''h,$$

so that

$$i = i_1, \quad i' = i_1 + i_2, \quad i'' = i_1 + i_2 + i_3,$$

then it is easily seen from the known properties of the solution of the problem of the moon's motion, that the lowest orders are as follows:—

(40) $|i - i'| + |i' - i''|$ for i', i'' even,

(41) $2 + |i - i'| + |i' - i''|$ for i', i'' odd,

(42) $2 + |i - i'|$ + smaller of $|i' - i'' \pm 1|$ for i' even, i'' odd and positive†,

(43) $4 + |i - i'|$ + smaller of $|i' - i'' \pm 1|$ for i' odd, i'' even and positive†.

The first expression (40) represents the orders for all terms in longitude and parallax or in x, y which are independent of $\frac{a}{a'}$; (41) for terms containing odd powers of $\frac{a}{a'}$ as a factor; (42) for terms in latitude or in z independent of $\frac{a}{a'}$, and (43) for those containing an odd power of $\frac{a}{a'}$ as a factor. Unity has been added in the expressions (42), (43) since an odd power of γ in R is always associated with an odd power of γ'', and conversely.

Method for finding the Periods to be Examined.

The lunar arguments l, g, h are first considered. These are divided into classes according to the multiple of l present; when the multiple is greater than 3, the value of p is so large that the coefficients are quickly seen to be insensible; in fact, there are none retained by the sieve with the multiple 3.

Consider any multiple $i'g + i''h$ of g, h. All the facts needed concerning a given multiple of g, h can be inserted on a sheet of squared paper by taking i', i'' as the coordinates and inserting in the squares: (1) the number of seconds in the argument; (2) the order with respect to e, γ of the coefficient as derived from the expressions (40), (41), (42), (43) associated with the argument; (3) the multiple of T associated with it in the greatest term of this order in the moon portions of R; that is, the term whose coefficient in these functions does not contain e' as a factor. The specimen given

* The results were worked out with l, g, h instead of with w_1, w_2, w_3.

† The positive sign is attached for convenience: the sign of the coefficient can always be arranged so that i'' is positive.

on page 47 includes all terms of the seventh order. It is evident that only two quadrants are required.

Similar charts were constructed for $l+i'g+i''h$, $2l+i'g+i''h$, ... but that it was not necessary to insert the values of s will presently appear.

For the planet arguments the table which follows was formed to give the periods of $pT-p_1V$ up to the first very long one, excluding periods greater than that of T. The last three* given are the arguments of very long periods: it is not necessary to go further, since all the periods greater than a year are either multiples of those given or are combinations of the last three with the others; higher multiples of T, V make the difference $|p-p_1|$ (which gives the order of the coefficient with respect to e', e'', γ'') so great that the terms would obviously become insensible if $s>1''$. We thus have all the periods of arguments independent of l, g, h. This particular table also illustrates the plan on which the sieve was worked.

Arg.	s	$\|N\|$	$\log\|C\|$	For Cal.
T	+3548	11·0	−·9 −·5	×
$12T-8V$	−3563	·03	−3·5	
$4T-3V$	−3110	14·	−·9 −·5	×
$9T-5V$	+3096	·04	no	
$9T-6V$	−2672	·3	−2·4	
$4T-2V$	+2657	2·4	−1·5	×
$T-V$	−2220	230·	+·6 −·5	×
$12T-7V$	+2204	·002	no	
$6T-4V$	−1782	2·3	−1·3	×
$7T-4V$	+1766	·3	−2·2	
$11T-7V$	−1344	·03	no	
$2T-V$	+1328	14·	−·5 −·5	×
$3T-2V$	−891	13·	−·3 −·5	×
$10T-6V$	+876	·03	−2·9	
$8T-5V$	−454	·25	−1·5	×
$5T-3V$	+438	2·3	−·7	×
$13T-8V$	−15	·004	−2·0	×
$26T-16V$	−30		no	
$39T-24V$	−45		no	

Formula: $\log C = 1{\cdot}5 - \log s + \log N_3$.

N_3	Max. of s
200	80000
20	8000
2	800
·2	80
·02	8
·002	1

−·5 to be added to $\log|C|$ when the multiples of T, $-V$ differ by less than 2.

* These three are multiples of the first of them, but it is convenient to have them written down.

Multiples of h \ Multiples of g	0	1	2	3	4	5	6	
7								
6	$6 - 6T$ -1145		$6 - T$ 38		$6 - 6T$ 1222		$6 - 6T$ 2405	
5	$6\ {}^{-4}_{-6}T$ -954	$7 - 5T$ -362	$6\ {}^{-4}_{-6}T$ 229	$7 - 5T$ 821	$6\ {}^{-4}_{-6}T$ 1413	$7 - 5T$ 2004		
4	$4 - 4T$ -763	$7\ {}^{-3}_{-5}T$ -171	$4 - 4T$ 420	$7\ {}^{-3}_{-5}T$ 1012	$4 - 4T$ 1604			
3	$4\ {}^{-2}_{-4}T$ -572	$5 - 3T$ 20	$4\ {}^{-2}_{-4}T$ 611	$5 - 3T$ 1203	$6\ {}^{-2}_{-4}T$ 1795			
2	$2 - 2T$ -382	$5\ {}^{-1}_{-3}T$ 210	$2 - 2T$ 801	$7\ {}^{-1}_{-3}T$ 1393	$6 - 2T$ 1985			
1	$2\ {}^{0}_{-2}T$ -191	$3 - T$ 401	$4\ {}^{0}_{-2}T$ 992	$7 - T$ 1584				
0		$5\ \mp T$ 592	$4\ \ 0T$ 1183					
-1		$5 + T$ 783	$6\ {}^{2}_{0}T$ 1374					
-2		$7\ {}^{3}_{1}T$ 974	$6\ \ 2T$ 1565					
-3		$7\ \ 3T$ 1164						
-4								

For arguments containing g, h but not l, we proceed according to the order given on the preceding page (for g, h), beginning with the lowest, which is 2. There are three terms of this order with arguments:

$$2g + 2h, \qquad 2h, \qquad h,$$

and motions

$$801'', \qquad -382'', \qquad -191''.$$

Taking the first, which arises principally from the terms $x^2 - y^2$, $2xy$ in R in the form $2\text{D} - 2l$, we examine by means of the second formula on page 44 all the terms which have values of s less than 3500, obtained by adding $\pm 801''$ to the values of s in the table given on page 46. For example, the argument

$$2g + 2h - 10T + 6V \text{ gives } s = -74.$$

To obtain N_3 we write it

$$2\text{D} - 2l - 8T + 6V,$$

where D is Delaunay's notation for the difference of the mean longitudes of the sun and moon. The value of N_3, obtained from the term in Newcomb's table corresponding to $-8T + 6V$, is extracted, and $\log C$ is then obtained. Before finding the terms of a given order, the maximum value of s needed for that order was found and the process was followed until the maximum value of s became less than $1''$.

For terms containing the first multiple of l, the most convenient one of very long period is taken, say $l + 16T - 18V$ in the case of Venus, for which $s = -13''$. This, after the maximum of s has been found, is combined with the motions in the table of arguments $pT - p_1 V$, giving a list of terms to be examined. The next first order term is $2\text{D} - l$, which is treated in the same manner; the starting term is $2\text{D} - l + 8T - 12V$, for which $s = -87''$. The same process is followed with every moon argument: it proceeds according to the orders of the coefficients with respect to e, γ, and is continued until an order is reached for which the maximum of s is less than $1''$.

The charts and tables which were constructed, together with the expressions constituting the sieve, permit one to review rapidly *all* the possible terms without any fear that some may be omitted. All the major planets from Mercury to Neptune were examined. For the minor planets a different method of treatment would be necessary: the principal effect is probably that of a circular ring of matter at the average mean distance: but the mass would be an exceedingly doubtful quantity.

It should be added that the numbers given in this section are only rough and ready approximations. Accurate values were afterwards used in the actual computations of those terms retained by the sieve.

SECTION V.

AUXILIARY NUMERICAL TABLES.

Epoch 1850·0.

Arg.	Daily motions of arguments		Longitudes at Epoch: Perigee	Longitudes at Epoch: Node
w_1	47434″·891	Mercury	75° 07′ 19″	46° 33′ 12″
w_2	400·923	Venus	129° 27′ 34″	75° 19′ 47″
w_3	− 190·772	Earth............	100° 21′ 40″	
Q	14732·420	Mars	333° 17′ 55″	48° 24′ 01″
V	5767·670	Jupiter	11° 54′ 27″	98° 55′ 58″
T	3548·193	Saturn	90° 06′ 40″	112° 20′ 51″
M	1886·518			
J	299·129			
S	120·455			

	Eccentricity	Inclination	Sine half inclin.	$\log \frac{a''}{a'}$	$\frac{m'}{m''}$
Moon e = ...	·10955		k ·044780		
„ e = ...	·054906		γ ·044887		
Earth.........	·016772				
Mercury ...	·205604	7° 00′ 07″	·061066	$\bar{1}$·5878216	6000000
Venus	·0068446	3° 23′ 35″·3	·0296063	$\bar{1}$·8593374	408000
Mars	·093261	1° 51′ 02″	·016149	·1828960	3093500
Jupiter	·048254	1° 18′ 42″	·011466	·7162374	1047·35
Saturn	·056061	2° 29′ 39″	·022	·9794957	3501·6

$$\alpha_1 = \frac{a}{a'} \cdot \frac{\text{diff. of masses of Earth and Moon}}{\text{sum} \quad ,, \quad\quad ,, \quad\quad ,,} = \cdot 0025053.$$

$a'A_p^i$. *Venus.*

i	$p=0$	$p=1$	$p=2$	$p=3$	$p=4$
0	2·386373	1·18898	2·01958	3·56717	7·14018
1	·942412	1·64375	1·97018	3·61813	7·15763
2	·527578	1·49718	2·23897	3·67099	7·24792
3	·323341	1·25777	2·38378	3·90485	7·38420
4	·206787	1·01817	2·37596	4·18715	7·66163
5	·135585	·806423	2·25093	4·40529	8·06778
6	·0903733	·629509	2·05234	4·50184	8·52231
7	·0609449	·486319	1·81709	4·46256	8·93131
8	·0414597	·372759	1·57213	4·30024	9·21822
9	·0283961	·283965	1·33525	4·04108	9·33621
10	·0195545	·215254	1·11704	3·71519	9·26837
11	·0135258	·162505	·922804	3·35128	9·02179
12	·00939037	·122264	·754294	2·97337	8·62021
13	·00653977	·0917207	·610989	2·60025	8·09624
14	·00456677	·0686356	·491048	2·24529	7·48568
15	·00319644	·0512491	·391967	1·91729	6·82294
16	·00224186	·0381937	·311003	1·62103	6·13881
17	·00157519	·0284159	·245453	1·35848	5·45869
18	·00110854	·0211093	·192801	1·12945	4·80230
19	·000781251	·0156602	·150799	·932354	4·18393
20	·000551305	·0116035	·117495	·764686	3·61295
21	·000389493	·00858810	·0912272	·623494	3·09465
22	·000275467	·00634988	·0706073	·505651	2·63101
23	·000195012	·00469065	·0544896	·408070	2·22155
24	·000138178	·00346202	·0419391	·327837	1·86399
25	·0000979878	·00255320	·0322001	·262283	1·55486
26	·0000695400	·00188159	·0246666	·209030	1·28999
27	·0000493857	·00138571	·0188559	·165996	1·06486
28	·0000350954	·00101988	·0143859	·131384	·87493
29	·0000249553	·000750191	·0109555	·103668	·71570
30	·0000177550	·000551514	·00832894	·0808952	·58391

$a'B_p^i$. *Venus.*

i	$p=0$	$p=1$	$p=2$	$p=3$	$p=4$
0	7·22787	40·7577	162·839	574·438	1889·97
1	6·41712	39·9376	162·009	572·558	1886·46
2	5·34305	37·4703	158·899	567·202	1875·65
3	4·30681	33·9489	153·053	557·873	1857·47
4	3·40299	29·9236	144·596	543·770	1831·35
5	2·65251	25·8036	134·045	524·376	1796·17
6	2·04714	21·8593	122·086	499·716	1750·67
7	1·56803	18·2494	109·417	470·353	1693·91
8	1·19391	15·0509	96·6472	437·240	1625·63
9	·904676	12·2853	84·2627	401·549	1546·42
10	·682782	9·93951	72·6128	364·502	1457·58
11	·513587	7·98023	61·9221	327·246	1361·05
12	·385215	6·36440	52·3111	290·774	1259·10
13	·288217	5·04589	43·8189	255·885	1154·16
14	·215180	3·97965	36·4253	223·167	1048·60
15	·160347	3·12409	30·0696	193·016	944·563
16	·119287	2·44218	24·6664	165·649	843·909
17	·0886081	1·90189	20·1175	141·143	748·121
18	·0657309	1·47604	16·3210	119·461	658·314
19	·0487009	1·14196	13·1767	100·483	575·238
20	·0360433	·880955	10·5906	84·0315	499·324
21	·0266486	·677817	8·47687	69·8952	430·720
22	·0196845	·520254	6·75904	57·8446	369·349
23	·0145280	·398422	5·37015	47·6466	314·956
24	·0107140	·304484	4·25252	39·0734	267·158
25	·00789548	·232245	3·35707	31·9104	225·485
26	·00591455	·176824	2·64251	25·9593	189·415
27	·00427940	·134402	2·07441	21·0408	158·405
28	·00314773	·101996	1·62429	16·9955	131·913
29	·00231406	·0772887	1·26880	13·6833	109·411
30	·00170032	·0584848	·988870	10·9828	90·4019

$a'C_p{}^i$. *Venus.*

i	$p=0$	$p=1$	$p=2$
0	44·8769	534·588	3768·39
1	43·6375	527·704	3741·20
2	40·5989	507·963	3660·39
3	36·5135	477·744	3529·34
4	31·9852	440·064	3354·28
5	27·4388	397·947	3143·75
6	23·1435	354·052	2907·47
7	19·2502	310·509	2655·40
8	15·8260	268·888	2396·86
9	12·8827	230·239	2139·95
10	10·3980	195·177	1891·28
11	8·33082	163·976	1655·86
12	6·63168	136·655	1437·18
13	5·24910	113·060	1237·38
14	4·13379	92·9238	1057·42
15	3·24075	75·9167	897·406
16	2·53032	61·6831	756·725
17	1·96836	49·8668	634·298
18	1·52609	40·1283	528·726
19	1·17959	32·1543	438·447
20	·909213	25·6635	361·821
21	·699011	20·4082	297·239
22	·536133	16·1740	243·145
23	·410305	12·7778	198·107
24	·313370	10·0649	160·803
25	·238882	7·90608	130·066
26	·181778	6·19418	104·853
27	·138097	4·84113	84·2627
28	·104749	3·77495	67·5142
29	·0793390	2·93721	53·9426
30	·0600106	2·28070	42·9841

Venus.

i	$a'A_0^i$
31	+ ·0000126389
32	900151
33	641390
34	457215
35	326061
36	232619
37	166018
38	118527
39	846489
40	604735
41	432154
42	308914
43	220879
44	157975
45	113014

i	$a'B_0^i/a$
31	+ ·00172640
32	126734
33	929960
34	682120
35	500139
36	366576
37	268590
38	196732
39	144055
40	105451
41	771713
42	564604
43	412972
44	301987
45	

i	$a'D_0$	$a'D_1$	$\frac{7}{4}a'Ǝ_0$
0	338·022	6140·05	4785·68
1	333·468	6086·23	4743·14
2	320·567	5928·97	4618·95
3	300·989	5679·86	4422·57
4	276·750	5355·68	4167·46
5	249·813	4975·76	3869·07
6	221·872	4559·78	3542·93
7	194·267	4126·04	3203·45
8	167·972	3690·36	2862·99
9	143·625	3265·59	2531·56
10	121·595	2861·50	2216·68
11	102·033	2484·89	1923·59
12	84·9390	2140·01	1655·50
13	70·2012	1828·94	1413·96
14	57·6434	1552·09	1199·19
15	47·0520	1308·57	1010·45
16	38·1990	1096·59	846·298
17	30·8580	913·798	704·860
18	24·8142	757·498	584·005
19	19·8702	624·874	481·532
20	15·8493	513·126	395·238
21	12·5965	419·566	323·039
22	9·97758	341·697	262·975
23	7·87850	277·235	213·283
24	6·20282	224·141	172·372
25	4·87014	180·612	138·849
26	3·81398	145·080	111·495
27	2·97966	116·193	89·267
28	2·32255	92·796	71·270
29	1·80647	73·913	56·751
30	1·40221	58·724	45·076

$a' \{(D+1)^2 - i^2\} A_p^i$. *Venus.*

i	$p=0$	$p=1$	$p=2$	$p=3$
0	9·99247	46·3548	178·7689	615·388
1	8·87162	46·3419	177·6337	613·923
2	7·38673	44·4156	175·2606	608·892
3	5·95414	40·9800	170·6122	600·643
4	4·70461	36·6645	163·2380	588·520
5	3·66708	32·0061	153·3097	571·636
6	2·83015	27·3902	141·3932	549·461
7	2·16779	23·0619	128·2057	522·054
8	1·65057	19·1572	114·4568	490·025
9	1·25071	15·7336	100·7589	454·380
10	·943940	12·7974	87·5890	416·333
11	·710030	10·3226	75·2842	377·131
12	·532556	8·26618	64·0535	337·940
13	·398458	6·57744	54·0018	299·757
14	·297484	5·20435	45·1531	263·374
15	·221679	4·09735	37·4736	229·369
16	·164913	3·21139	30·8897	198·119
17	·122500	2·50685	25·3054	169·824
18	·0908724	1·94975	20·6139	144·540
19	·0673285	1·51142	16·7052	122·211
20	·0498295	1·16808	13·4733	102·699
21	·0368414	·900236	10·8190	85·8106
22	·0272136	·692034	8·65228	71·3175
23	·0200849	·530730	6·89346	58·9776
24	·0148120	·406135	5·47294	48·5458
25	·0109154	·310161	4·33096	39·7849
26	·00803856	·236420	3·41683	32·4717
27	·00591623	·179894	2·68796	26·4008
28	·00435171	·136657	2·10891	21·3874
29	·00319917	·103652	1·65045	17·2666
30	·00235068	·078504	1·28460	13·8970

$a' \{(D+1)^2 - (i+1)^2\} B_p{}^i$, i positive. *Venus.*

i	$p=0$	$p=1$	$p=2$
0	447·951	5197·28	36045·4
1	424·582	5055·43	35471·2
2	387·464	4804·84	34419·0
3	343·350	4470·38	32933·5
4	297·290	4080·18	31081·8
5	252·663	3660·99	28946·7
6	211·488	3235·50	26618·6
7	174·796	2821·32	24186·4
8	142·934	2431·00	21731·0
9	115·818	2072·54	19321·6
10	93·1103	1750·22	17012·9
11	74·3419	1465·46	14845·2
12	58·9993	1217·63	12845·7
13	46·5732	1004·69	11029·4
14	36·5893	823·755	9401·82
15	28·6228	671·520	7960·90
16	22·3046	544·536	6698·96
17	17·3203	439·428	5604·44
18	13·4070	353·028	4663·47
19	10·3476	282·448	3860·91
20	7·96501	225·117	3181·42
21	6·11593	178·788	2609·92
22	4·68543	141·525	2132·23
23	3·58195	111·684	1735·17
24	2·73298	87·8809	1406·89
25	2·08143	68·9647	1136·77
26	1·58250	53·9832	915·517
27	1·20125	42·1554	735·058
28	·910479	32·8454	588·447
29	·689121	25·5372	469·778
30	·520888	19·8153	374·056

$a' \{(D+1)^2 - (i+1)^2\} B_p^i$, i negative. *Venus.*

$-i$	$p=0$	$p=1$	$p=2$
0	447·951	5197·28	36045·4
1	450·251	5215·18	36119·2
2	430·208	5104·60	35690·2
3	395·032	4877·77	34770·1
4	351·738	4558·96	33395·3
5	305·713	4177·06	31627·6
6	260·619	3760·12	29548·7
7	218·701	3332·30	27250·1
8	181·139	2912·63	24823·7
9	148·386	2514·81	22355·1
10	120·4216	2147·80	19917·4
11	96·9397	1816·59	17569·8
12	77·4986	1523·12	15356·6
13	61·5605	1267·08	13308·0
14	48·6395	1046·62	11441·6
15	38·2436	858·965	9765·07
16	29·9390	700·836	8277·61
17	23·3457	568·756	6972·43
18	18·1396	459·303	5838·58
19	14·0489	369·237	4862·34
20	10·8485	295·593	4028·67
21	8·35441	235·725	3321·98
22	6·41766	187·307	2727·03
23	4·91853	148·339	2229·22
24	3·76152	117·111	1815·13
25	2·87098	92·1892	1472·47
26	2·18721	72·3729	1190·34
27	1·66343	56·6708	959·094
28	1·26303	44·2685	770·368
29	·957552	34·5027	616·958
30	·724926	26·8335	492·720

$a' \{(D+1)^2 - i^2\} C_p^i$. *Venus.*

i	$p=0$	$p=1$
0	9185·44	163970·
1	9065·52	162567·
2	8722·88	158461·
3	8199·80	151939·
4	7548·99	143425·
5	6822·81	133414·
6	6067·08	122420·
7	5318·32	110922·
8	4603·33	99340·4
9	3940·00	88020·2
10	3338·69	77225·0
11	2803·95	67142·4
12	2336·00	57890·6
13	1932·08	49530·5
14	1587·52	42077·2
15	1296·63	35510·7
16	1053·26	29786·5
17	851·310	24843·7
18	684·906	20611·9
19	548·705	17016·9
20	437·852	13984·5
21	348·138	11443·0
22	275·855	9325·72
23	217·905	7571·41
24	171·617	6125·22
25	134·788	4938·65
26	105·589	3969·32
27	82·5145	3180·71
28	64·3347	2541·55
29	50·0521	2025·38
30	38·8608	1609·93

Jupiter.

i	$a''A_0^i$	$a''A_1^i$	$a''A_2^i$	$a''A_3^i$
0	2·018865	·0385392	·0209301	·00174900
1	·194930	·200512	·0086394	·00333975
2	·0281439	·0571840	·0304292	·00192985
3	·00451137	·0136850	·0140702	·00522212
4	·000759075	·00306243	·00467341	·00325231
5	·000131349	·000661353	·00133908	·00137257
6	·0000231472	·000139706		

i	$a''B_0^i$	$a''B_1^i$	$a''B_2^i$	$a''B_3^i$
0	·418302	·490177	·116415	·0540404
1	·118939	·255013	·163696	·0402207
2	·0284421	·0891538	·0991362	·0473610
3	·00636279	·0262769	·0419796	·0325014
4	·00137387	·00704369	·0147155	·0160281
5	·000290196	·00177743	·00459230	·00648290
6	·0000603839	·000430147	·00132477	·00230235

i	$a''C_0^i$	$a''C_1^i$
0	·0930465	·228991
1	·0418422	·139507
2	·0137542	·0589828
3	·00391946	·0206353
4	·00102857	·00642907
5	·000255794	·00185218
6	·0000612487	·000504318

i	$a''D_0^i$
0	·0219945
1	·0128071
2	·00525764
3	·00180378
4	·000554293
5	·000158071
6	·0000427049

Mars.

i	$a''A_0^i$	$a''B_0^i$
0	2·29114	4·49988
1	·804571	3·75928
2	·405593	2·89074
3	·224607	2·13691
4	·129983	1·54309
5	·0771836	1·09704
6	·0466132	·771258
7	·0284904	·537681
8	·0175701	·372392
9	·0109110	·256560
10	·00681334	·175994
11	·00427387	·120293
12	·00269104	·0819691
13	·00169982	·0557079
14	·00107664	·0377737
15	·000683546	·0255619
16	·000434874	·0172673
17	·000277173	·0116459
18	·000176948	·00784346
19	·000113128	·00527580
20	·0000724205	·00354460
21	·0000464159	·00237898
22	·0000297811	·00159513
23	·0000191267	·00106861
24	·0000122950	·000715304
25	·00000791006	·000478450
26	·00000509286	·000319801
27	·00000328134	·000213621
28	·00000211557	·000142609
29	·00000136480	·0000951489
30	·000000880963	·0000634503

i	$a''A_1^i$	$a''A_2^i$	$a''B_1^i$	$a''B_2^i$
0	·805995	1·07364	19·0483	55·6199
1	1·228084	1·02186	18·4080	55·2557
2	1·050895	1·23434	16·5922	53·5762
3	·814457	1·30473	14·2044	50·3183
4	·604237	1·24412	11·7054	45·7478
5	·437186	1·10620	9·37056	40·3654
6	·311182	·936535	7·33356	34·6859

i	$a''A_1^i$	$a''A_2^i$	$a''A_3^i$
20	·00150106	·0148967	·0943680
23	·000453886	·00518273	·0379559
26	·000136149	·00175826	·0146198

i	$a''B_1^i$	$a''B_2^i$
20	·0827974	·940514
23	·0281560	·361317
26	·00938295	·134286

i	$a''C_0^i$
0	16·3697
1	15·5930
2	13·8635
3	11·7387
4	9·58972
5	7·62374
6	5·93292
20	·0649997
23	·0220512
26	·00733458

Mercury.

i	$a'A_0^i$	$a'A_1^i$	$a'A_2^i$	$a'A_3^i$	$a'A_4^i$
0	2·081981	·179759	·1252855	·0440444	·0241364
1	·411140	·464376	·0914646	·0528043	·0227489
2	·120179	·257705	·167519	·0476530	·0255266
3	·0389015	·122616	·138789	·0714497	·0252692
4	·0132039	·0548838	·0890266	·0719791	·0335508
5	·00460655	·0237688	·0502928	·0563902	·0371389
6	·00163624	·0100826	·0263323	·0379722	·0333614
7	·000588587	·00421654	·0131072	·0231604	·0257195
8	·000213722	·00174517	·00629294	·0131806	·0177725

i	$a'B_0^i$	$a'B_1^i$	$a'B_2^i$	$a'B_3^i$	$a'B_4^i$
0	1·111682	1·977295	1·772252	1·607113	1·259825
1	·610090	1·626068	1·834149	1·536827	1·265314
2	·289401	1·048542	1·589323	1·501504	1·214326
3	·129374	·595249	1·164074	1·346214	1·163209
4	·0559922	·312846	·756591	1·072807	1·062710
5	·0237426	·156179	·451861	·770384	·894799
6	·00992677	·0751578	·253663	·508904	·691321
7	·00410781	·0351878	·135874	·314718	·494563
8	·00168656	·0161264	·0701629	·184660	·331247

i	$a'C_0^i$	$a'C_1^i$	$a'C_2^i$	$a'D_0^i$
0	·768937	3·01626	5·98351	·619377
1	·586193	2·64675	5·68166	·535436
2	·362211	1·93221	4·76074	·384900
3	·200324	1·24869	3·56256	·245667
4	·103463	·741711	2·43259	·144511
5	·0510119	·414433	1·54498	·0801264
6	·0243209	·221116	·926391	·0424868
7	·0113048	·113801	·530331	·0217562
8	·00515116	·0569059	·292314	·0108334

Values of M_i and of their derivatives. Arguments in the Moon portions of R.

Terms containing nl' in the argument are to be multiplied by e'^{n}.

For brevity, $\frac{n}{a^2}\frac{d}{dn}(M_i a^2)$ has been printed $n\frac{d}{dn}M_i$.

Argument	0	$\pm l'$	$\pm 2l'$	$\pm 3l'$
M_1	$+ \cdot 99276$	$+ 1\cdot 0005$	$+ 1\cdot 259$	$+ 1\cdot 65$
M_2	$- \cdot 0140 \times 2$	$- \cdot 0835$	$- \cdot 201$	
M_3	$0\cdot$	$\mp \cdot 0107$	$\mp \cdot 040$	
$n\frac{d}{dn}M_1$	$- 1\cdot 3240$	$- 1\cdot 3478$	$- 1\cdot 701$	$- 2\cdot 21$
$n\frac{d}{dn}M_2$	$+ \cdot 0540 \times 2$	$+ \cdot 3245$	$+ \cdot 852$	$+ 2\cdot 0$
$n\frac{d}{dn}M_3$	$0\cdot$	$\pm \cdot 0509$	$\pm \cdot 302$	$0\cdot$
$\frac{d}{de}M_1$	$+ \cdot 0842$	$+ \cdot 1009$	$+ \cdot 141$	
$\frac{d}{de}M_2$	$+ \cdot 0558 \times 2$	$+ \cdot 2474$	$+ \cdot 377$	
$\frac{d}{de}M_3$	$0\cdot$	$\mp \cdot 0368$	$\pm \cdot 029$	
$\frac{d}{dk}M_1$	$- \cdot 5390$			
$\frac{d}{dk}M_2$	$+ \cdot 0163 \times 2$			

Argument	$\pm l$	$\pm(l+l')$	$\pm(l-l')$	$\pm(l+2l')$	$\pm(l-2l')$
$\frac{1}{e}M_1$	$- \cdot 49080$	$- \cdot 3452$	$- \cdot 6736$	$- \cdot 370$	$- \cdot 978$
$\frac{1}{e}M_2$	$- \cdot 2862$	$- \cdot 421$	$- \cdot 799$	$- \cdot 73$	$- 1\cdot 10$
$\frac{1}{e}M_3$	$\mp \cdot 3108$	$\mp \cdot 477$	$\mp \cdot 889$	$\mp \cdot 83$	$\mp 1\cdot 34$
$\frac{n}{e}\frac{d}{dn}M_1$	$+ \cdot 5926$	$+ \cdot 1813$	$+ 1\cdot 1544$	$+ \cdot 065$	$+ 1\cdot 904$
$\frac{n}{e}\frac{d}{dn}M_2$	$+ \cdot 7523$	$+ 1\cdot 1117$	$+ 2\cdot 5564$	$+ 2\cdot 151$	$+ 3\cdot 595$
$\frac{n}{e}\frac{d}{dn}M_3$	$\pm \cdot 8423$	$\pm 1\cdot 3085$	$\pm 2\cdot 9124$	$\pm 2\cdot 509$	$\pm 4\cdot 643$

$e\frac{d}{de}M_i = M_i$ with a sufficient approximation

Argument	$\pm 2l$	$\pm(2l+l')$	$\pm(2l-l')$
$\frac{1}{e^2} M_1$	− ·0618	− ·0247	− ·1078
$\frac{1}{e^2} M_2$	− ·2069	− ·24	− ·73
$\frac{1}{e^2} M_3$	∓ ·2065	∓ ·24	∓ ·73
$\frac{n}{e^2}\frac{d}{dn} M_1$	+ ·0986	− ·001	+ ·230
$\frac{n}{e^2}\frac{d}{dn} M_2$	+ ·5453	+ ·586	+ 2·614
$\frac{n}{e^2}\frac{d}{dn} M_3$	± ·5441	± ·582	± 2·604

$$e\frac{d}{de} M_i = 2M_i \text{ with a sufficient approximation}$$

Argument	$\pm 2D$	$\pm(2D+l')$	$\pm(2D-l')$	$\pm(2D+2l')$	$\pm(2D-2l')$
M_1	− ·00701	− ·00291	− ·03693	− ·0042	− ·1265
M_2	+ ·9870	− 1·1735	+ 3·1571	+ ·0826	+ 7·1584
M_3	± ·9868	∓ 1·1739	± 3·1561	± ·0818	± 7·1534
$\frac{d}{dn} M_1$	+ ·0254	+ ·0103	+ ·1394	+ ·017	+ ·500
$\frac{d}{dn} M_2$	− 1·3144	+ 1·6697	− 4·3241	− ·239	− 9·928
$\frac{d}{dn} M_3$	∓ 1·3136	± 1·6709	∓ 4·3169	∓ ·239	∓ 9·892
$\frac{d}{de} M_1$	− ·0052	− ·0013	− ·0214	+ ·009	− ·061
$\frac{d}{de} M_2$	− ·1445	+ ·625	− ·4104	+ ·004	− ·838
$\frac{d}{de} M_3$	∓ ·1445	± ·625	∓ ·4100	± ·004	∓ ·836

Argument	$\pm(l-2D)$	$\pm(l-2D+l')$	$\pm(l-2D-l')$	$\pm(l-2D+2l')$	$\pm(l-2D-2l')$
$\frac{1}{e} M_1$	− ·08582	− ·2948	− ·0520	− ·7197	− ·0716
$\frac{1}{e} M_2$	− 1·4897	− 4·321	+ 1·188	− 9·11	− ·03
$\frac{1}{e} M_3$	± 1·4944	± 4·349	∓ 1·178	± 9·21	± ·05
$\frac{n}{e}\frac{d}{dn} M_1$	+ ·2220	+ ·7383	+ ·2751	+ 1·785	− ·425
$\frac{n}{e}\frac{d}{dn} M_2$	+ 1·8274	+ 4·797	− ·501	+ 9·01	+ ·24
$\frac{n}{e}\frac{d}{dn} M_3$	∓ 1·8518	∓ 4·941	± ·440	∓ 9·67	∓ ·31

$$e\frac{d}{de} M_i = M_i$$

Argument	$\pm(2l-2D)$	$\pm(2l-2D+l')$	$\pm(2l-2D-l')$
$\frac{1}{e^2} M_1$	$+$ ·0758	$+$ ·2552	$+$ ·0785
$\frac{1}{e^2} M_2$	$\mp$ 0107	[illegible]	10
$\frac{1}{e^2} M_3$	$\mp$ ·5969	$\mp$ 1·53	$\pm$ ·32
$\frac{n}{e^2}\frac{d}{dn} M_1$	$-$ ·2228	$-$ ·663	$-$ ·400
$\frac{n}{e^2}\frac{d}{dn} M_2$	$-$ ·8064	$-$ 1·81	$-$ ·93
$\frac{n}{e^2}\frac{d}{dn} M_3$	$\pm$ ·6010	$\pm$ ·93	$\pm$ ·27

$$e\frac{d}{de} M_i = 2M_i$$

Argument	$\pm 2F$	$\pm(2F+l')$	$\pm(2F-l')$
$\frac{1}{k^2} M_1$	$+$ 2·9901	$+$ 2·8805	$+$ 3·0882
$\frac{1}{k^2} M_2$	$-$ ·2001	$-$ ·25	$-$ ·37
$\frac{1}{k^2} M_3$	$\mp$ ·1717	$\mp$ ·19	$\mp$ ·27
$\frac{n}{k^2}\frac{d}{dn} M_1$	$-$ 2·9549	$-$ 2·937	$-$ 2·939
$\frac{n}{k^2}\frac{d}{dn} M_2$	$+$ ·3785	$+$ ·509	$+$ ·645
$\frac{n}{k^2}\frac{d}{dn} M_3$	$\pm$ ·2873	$\pm$ ·275	$\pm$ ·349

$$k\frac{d}{dk} M_i = 2M_i$$

Argument	$\pm(2F-2D)$	$\pm(2F-2D+l')$	$\pm(2F-2D-l')$
$\frac{1}{k^2} M_1$	$-$ ·2062	$-$ ·8051	$-$ ·1653
$\frac{1}{k^2} M_2$	$+$ 1·9756	$+$ 6·22	$-$ 2·31
$\frac{1}{k^2} M_3$	$\mp$ 1·9690	$\mp$ 6·18	$\pm$ 2·31
$\frac{n}{k^2}\frac{d}{dn} M_1$	$+$ ·4637	$+$ 1·967	$-$ ·319
$\frac{n}{k^2}\frac{d}{dn} M_2$	$-$ 1·9243	$-$ 6·282	$+$ 2·583
$\frac{n}{k^2}\frac{d}{dn} M_3$	$\pm$ 1·9019	$\pm$ 6·164	$\mp$ 2·605

$$k\frac{d}{dk} M_i = 2M_i$$

Argument	$2l+2g+h-h''$	$2l+2g+h-h''+l'$	$2l+2g+h-h''-l'$	
$\frac{1}{\mathrm{k}} M_4$	$+1{\cdot}0000$	$+\ {\cdot}8917$	$+1{\cdot}1175$	$\mathrm{k}\frac{d}{d\mathrm{k}} M_4 = M_4$
$\frac{n}{\mathrm{k}}\frac{d}{dn} M_4$	$-1{\cdot}1646$	$-1{\cdot}0207$	$-1{\cdot}3401$	

Argument	$h-h''$	$h-h''+l'$	$h-h''-l'$	
$\frac{1}{\mathrm{k}} M_4$	$-1{\cdot}0004$	$-1{\cdot}0838$	$-\ {\cdot}9285$	,,
$\frac{n}{\mathrm{k}}\frac{d}{dn} M_4$	$+1{\cdot}16399$	$+1{\cdot}3385$	$+1{\cdot}0256$	

Argument	$2T-h-h''$	$2T-h-h''+l'$	$2T-h-h''-l'$	
$\frac{1}{\mathrm{k}} M_4$	$-\ {\cdot}0457$	$-\ {\cdot}1887$	$-\ {\cdot}0041$	,,
$\frac{n}{\mathrm{k}}\frac{d}{dn} M_4$	$+\ {\cdot}1213$	$+\ {\cdot}7669$	$-\ {\cdot}0592$	

Argument	$2D+h-h''$	$2D+h-h''+l'$	$2D+h-h''-l'$	
$\frac{1}{\mathrm{k}} M_4$	$+\ {\cdot}0355$	$+\ {\cdot}1421$	$-\ {\cdot}0078$	,,
$\frac{n}{\mathrm{k}}\frac{d}{dn} M_4$	$-\ {\cdot}0857$	$-\ {\cdot}3682$	$+\ {\cdot}0227$	

Argument	D		$4D-3l$		$4D-4F$		$l-2F$		$3l-2D$	
M_1	$+{\cdot}1132$		$-{\cdot}002258$		${\cdot}00$		$-3{\cdot}497$		$+{\cdot}02080$	
M_2+M_3	$+{\cdot}7182$	α_1	$-{\cdot}04854$	e^3	$-{\cdot}0334$	k^4	$-\ {\cdot}3876$	ek^2	$+{\cdot}04856$	e^3
M_2-M_3	$-{\cdot}0180$		$-{\cdot}00066$		$-{\cdot}0004$		$+\ {\cdot}9779$		$-{\cdot}07420$	

Argument	$4D-l-2F$		$D-2l$		$3D-2F$		$l-D$	
M_1	$+{\cdot}01108$		$-{\cdot}1387$		$+\ {\cdot}0951$		$+{\cdot}00396$	
M_2+M_3	$+{\cdot}02282$	ek^2	$+{\cdot}048$	$e^2\alpha_1$	$+1{\cdot}960$	$k^2\alpha_1$	$-{\cdot}04627$	$e\alpha_1$
M_2-M_3	$-{\cdot}00056$		$+{\cdot}310$		$-\ {\cdot}006$		$-{\cdot}58098$	

Argument	$l+2g+h-h''$	$l+h-h''$	$-2l-2g+h-h''$	$3g+4h-h''-2T$
M_4	$-1{\cdot}49ek$	$+{\cdot}49ek$	$-{\cdot}989k^3$	$-{\cdot}0154e^3k$

Argument	D	$D-l'$		$D+l'$		$l-D$		$l+l'-D$	
M_6+M_7	$+1{\cdot}972$	$+5{\cdot}143$		$+{\cdot}815$		$-\ {\cdot}4620$		$-1{\cdot}38$	
M_6-M_7	$-\ {\cdot}0280$	$-\ {\cdot}134$	e'	$-{\cdot}056$	e'	$-2{\cdot}465$	e	$-5{\cdot}68$	ee'
M_8+M_9	$-\ {\cdot}0139$	$-\ {\cdot}058$		$-{\cdot}389$		$+\ {\cdot}0051$		$+\ {\cdot}023$	
M_8-M_9	$-\ {\cdot}000$	$+\ {\cdot}0009$		$+{\cdot}0005$		$+\ {\cdot}0227$		$+\ {\cdot}089$	

Argument	$3D-2F$		$3D-2F-l'$	
M_6+M_7	$-\ {\cdot}326$		$+\ {\cdot}62$	
M_6-M_7	$+\ {\cdot}012$	k^2	${\cdot}0$	k^2e'
M_8+M_9	$+1{\cdot}959$		$-3{\cdot}48$	
M_8-M_9	${\cdot}000$		${\cdot}0$	

Extract from the computations for illustration of the method.

$$\phi = (l) + (29T - 26V - 3\varpi') \qquad \text{factor } e\left(\frac{e'}{2}\right)^3.$$
$$\text{Leverrier } (270)^i = (112),\ i = 26\ ;\ K^i = A^i$$

$$\text{Lev. coef.} = \left[\tfrac{1}{6}(27 + 65i + 42i^2 + 8i^3) A_0^i + \tfrac{1}{2}(17 + 17i + 4i^2) A_1^i + (5 + 2i) A_2^i + A_3^i\right]\left(\frac{e'}{2}\right)^3$$

$$P = 28453 A_0 + 1581{\cdot}5 A_1 + 57 A_2 + A_3 \qquad \text{for } i = 26$$

$$DP = 30034{\cdot}5 A_1 + 3277 A_2 + 174 A_3 + 4A_4 \qquad [DK_p = (p+1) K_{p+1} + pK_p]$$

From the tables for Venus $P = +6{\cdot}568$, $DP = +178{\cdot}87$, $\{(D+1)^2 - i^2\} P = 829{\cdot}8 = P_1$

Factor $\left(\frac{e'}{2}\right)^3$ $\quad P_3 = 26(D+2) P = +4992$, $P_2 = \tfrac{1}{2} P_1 + (D+2) P + (26^2 - 1) P = +5041$

$s = -27{\cdot}85$

	log		log
$\frac{1}{s^2}$	$+3{\cdot}1104$	$\frac{1}{s}$	$-\bar{2}{\cdot}5552$
λ_1	$-\ {\cdot}0063$		
f	$6{\cdot}6829$	f'	$2{\cdot}0068$
e	$\bar{1}{\cdot}0396$	e	$-\bar{1}{\cdot}0396$
prod.	$-2{\cdot}8392$		$-\bar{1}{\cdot}6016$
$\left(\frac{e'}{2}\right)^3$	$\bar{7}{\cdot}7707$		insen.
fac.	$-\bar{4}{\cdot}6099$		

$$\delta w_1 = \frac{\lambda_1 f A}{s^2} + \frac{f' A_1}{s},\quad \frac{f' A_1}{s} \text{ insensible}$$

P_1	$+2{\cdot}9190$	$\frac{1}{2}(P_2 + P_3)$	$+3{\cdot}7003$	$\frac{1}{2}(P_2 - P_3)$	$+1{\cdot}3892$
M_1	$-1{\cdot}6909$	$M_2 - M_3$	$+\bar{2}{\cdot}3909$	$M_2 + M_3$	$-\bar{1}{\cdot}7760$
fac.	$-\bar{4}{\cdot}6099$	fac.	$-\bar{4}{\cdot}6099$	fac.	$-\bar{4}{\cdot}6099$
	$+\bar{1}{\cdot}2198$		$-\bar{2}{\cdot}7011$		$+\bar{3}{\cdot}7751$
	$+0''{\cdot}166$		$-0''{\cdot}050$		$+0''{\cdot}006$

Portion in this term depending on

$$e\left(\frac{e'}{2}\right)^3 \gamma''^2.$$

All found from the tables except

$$[(D+1)^2 - 26^2] B_3^{27},$$

which is only needed within 50°/₀ of its true value. It can be computed from the general formulae, but the following estimate gives it with sufficient accuracy :

			P_1
$[(D+1)^2 - 26^2]$	E_0^{26}	$1{\cdot}8724$	$-\ 53270$
,,	E_1	$62{\cdot}82$	$-\ 99360$
,,	E_2	$1047{\cdot}9$	$-\ 59730$
,,	E_3 (est.)	$12000{\cdot}$	$-\ 12000$
			-224000

etc.

Replace A_p^i by $-\gamma''^2 E_p^i = -\frac{\gamma''^2}{2}(B_p^{i-1} + B_p^{i+1})$

$$= -\gamma''^2 (iA_p^i + B_p^{i+1})$$

The three parts are

$P_1 M_1$, $\tfrac{1}{2}(P_2 + P_3)(M_2 - M_3)$, $\tfrac{1}{2}(P_2 - P_3)(M_2 + M_3)$

P	$-5{\cdot}3502$	$-5{\cdot}7466$	$-4{\cdot}0055$
M	$-\bar{1}{\cdot}6909$	$+\bar{2}{\cdot}3909$	$-\bar{1}{\cdot}7760$
fac.	$-\bar{4}{\cdot}6099$	$-\bar{4}{\cdot}6099$	$-\bar{4}{\cdot}6099$
γ''^2	$\bar{4}{\cdot}9429$	$\bar{4}{\cdot}9429$	$\bar{4}{\cdot}9429$
	$-\bar{2}{\cdot}5939$	$+\bar{3}{\cdot}6903$	$-\bar{3}{\cdot}3343$
	$-0''{\cdot}039$	$+0''{\cdot}005$	$-0''{\cdot}002$

$$\phi = (l) + (29T - 26V - 3\varpi'), \qquad \text{factor e}\left(\frac{e'}{2}\right)^5.$$

Portion in this term with factor $\left(\frac{e'}{2}\right)^5$.

Levr. coef. $= (272)^i = (114)$, $i = 26$, $K^i = A^i$

$$(114) = \tfrac{1}{24}(243 - 858i - 1317i^2 - 926i^3 - 288i^4 - 32i^5)A_0^i$$
$$+ \tfrac{1}{24}(93 - 416i - 741i^2 - 344i^3 - 48i^4)A_1^i + \tfrac{1}{6}(309 + 163i - 6i^2 - 8i^3)A_2^i$$
$$+ \tfrac{1}{2}(123 + 51i + 4i^2)A_3^i + (29 + 6i)A_4^i + 5A_5^i$$

$i = 26$		P_1
$[(D+1)^2 - i^2]A_0^i +$	·008038	− 177200
,, $A_1^i +$	·2364	− 280300
,, $A_2^i +$	3·417	− 79800
,, $A_3^i +$	32·47	+ 67500
,, $A_4^i +$	270· (est.)	+ 50000
,, $A_5^i +$	2000· (est.)	+ 10000
		$-410000\left(\frac{e'}{2}\right)^5$.

Resulting coefficients.

From P_1	$-0''\cdot006$	$(-\frac{1}{28}$ of prin. pt.)
,, $\frac{P_2+P_3}{2}$	$+0''\cdot002$	
,, $\frac{P_2-P_3}{2}$	$-0''\cdot000$	

Portion with factor $\text{e}\left(\frac{e'}{2}\right)^3 . \left(\frac{e''}{2}\right)^2$.

Levr. coef. is approx. 4 times that for $\left(\frac{e'}{2}\right)^5$, and $\frac{e''^2}{e'^2} = \frac{1}{6}$ approx.; therefore take $\frac{2}{3}$ of coef. for $\left(\frac{e'}{2}\right)^5$. We obtain

from P_1	$-0''\cdot004$
,, $\frac{P_2+P_3}{2}$	$+0''\cdot001$

$\phi = (l + l') + (28T - 26V - 2\varpi')$, factor $ee'.\left(\frac{e'}{2}\right)^2$, $\left(\text{term } ee'\frac{e'^2}{2}\gamma''^2 \text{ est.} = \frac{1}{6} \text{ of principal part.}\right)$

Same method gives

from P_1	$+0''\cdot008$
,, $\frac{P_2+P_3}{2}$	$-0''\cdot007$
,, $\frac{P_2-P_3}{2}$	·000

Other terms insensible.

Summary for argument $(l) + (29T - 26V - 3\varpi')$

Factor	From P_1,	$\frac{P_2+P_3}{2}$,	$\frac{P_2-P_3}{2}$
$\text{e}\left(\frac{e'}{2}\right)^3$	$+0''\cdot166$	$-0''\cdot050$	$+0''\cdot006$
$\text{e}\left(\frac{e'}{2}\right)^3\gamma''^2$	− 39	+ 5	− 2
$\text{e}\left(\frac{e'}{2}\right)^5$	− 6	+ 2	
$\text{e}\left(\frac{e'}{2}\right)^3\left(\frac{e''}{2}\right)^2$	− 4	+ 1	
$ee'\left(\frac{e'}{2}\right)^2(1, \gamma''^2)$	+ 7	− 7	
	$+0''\cdot124$	$-0''\cdot049$	$+0''\cdot004 = +0''\cdot079$

The terms and coefficients retained and calculated by the sieve.

Argument	$\lvert s \rvert$	$\log \lvert C \rvert$
T	3548	$-1{\cdot}2$
$4T-3V$	3110	$-1{\cdot}4$
$4T-2V$	2656	$-1{\cdot}5$
$T-V$	2220	$+\ {\cdot}1$
$6T-4V$	1782	$-1{\cdot}3$
$2T-V$	1328	$-1{\cdot}0$
$3T-2V$	891	$-1{\cdot}8$
$8T-5V$	453	$-1{\cdot}5$
$5T-3V$	438	$-\ {\cdot}7$
$13T-8V$	15	$-2{\cdot}0$
$2h-8T+5V$	71	$-\ {\cdot}3$
„ $-5T+3V$	820	$-1{\cdot}7$
$2g+2h-10T+6V$	74	$-1{\cdot}1$
„ $-5T+3V$	364	$-1{\cdot}0$
$g+h-5T+3V$	37	$-\ {\cdot}3$
$h-6T+4V$	1592	$-1{\cdot}8$
$h-3T+2V$	700	$-1{\cdot}8$
$h-8T+5V$	263	$-1{\cdot}0$
$h+5T-3V$	247	$-1{\cdot}6$
$h-5T+3V$	629	$-\ {\cdot}9$
$l+3T-10V$	1·85	0·0
$l+16T-18V$	13	$+1{\cdot}7$
$l+21(T-V)$	425	$-\ {\cdot}3$
$l+24T-23V$	467	$-\ {\cdot}9$
$l+29T-26V$	28	$+\ {\cdot}1$
$l+22(T-V)$	1795	$-1{\cdot}5$
$(2D-l)+8T-12V$	87	$-1{\cdot}2$
„ $+21T-20V$	102	$+1{\cdot}0$
„ $+13T-15V$	351	$-\ {\cdot}7$
„ $+26T-23V$	336	$-1{\cdot}3$
„ $+16T-17V$	541	$-1{\cdot}2$
„ $+18T-18V$	789	$-\ {\cdot}1$
„ $+24T-23V$	993	$-1{\cdot}7$
„ $+23T-21V$	1226	$-1{\cdot}7$
„ $+19T-19V$	1431	$-1{\cdot}0$

Argument	$\|s\|$	$\log\|C\|$
$(2D-l)+20T-19V$	2117	−1·5
„ $+22T-21V$	2321	−1·7
„ $+17T-17V$	3009	−1·3
„ $+20T-20V$	3651	−1·7
$D+12T-15V$	50	−1·6
$D+25T-23V$	65	− ·4
$D+20T-20V$	505	−1·1
$3D-2F+19T-18V$	6·4	+ ·1
$F+24T-23V$	125	−1·6
$2D-F+18T-18V$	197	−1·8
$4D-l-2F+15T-15V$	32	+ ·9
$3D-F-l+14T-13V$	162	− ·2
$2D-2F+l+23T-21V$	43	− ·9
„ $+18T-18V$	395	−1·6
$D+l-F+22T-21V$	234	−1·3
„ $+17T-18V$	204	− ·7
$D-l+F+20T-20V$	88	− ·5
$3l-2D+24T-24V$	62	− ·7
$4D-3l+18T-17V$	232	−1·3
$2l-D+20T-21V$	24	−1·7
$3D-2F+19T-18V$	6·4	+ ·1
$3D-2l+17T-17V$	138	−1·5
$4D-2F-3l-17T+21V$	4·6	−1·4
$D+l-3F-26T+25V$	18	−1·5
$3D+F-3l+25T-22V$	0·07	+2·3
$2D+34T-36V$	775	−1·7
„ $+37T-38V$	116	− ·5
„ $+42T-41V$	322	−1·1
„ $+29T-33V$	338	−1·5
$2l+32T-36V$	26	−1·7
$D-3F-42T+43V$	4·45	−1·2
$3D+F-2l+28T-32V$	1·70	−1·1
$2D-l+5T-4Q$	449	−1·3
„ $+T-3Q$	90	− ·2
$l+3T-4Q$	1251	−1·6
$2D+8T-8Q$	1700	−1·3
„ $+4T-7Q$	1161	−1·2

Argument	$\lvert s \rvert$	$\log \lvert C \rvert$
$2\mathrm{D} - \mathrm{F} + l + 2T - 3Q$	100	− ·9
$2\mathrm{F} - l + 3T - 4Q$	68	− 1·1
$T - J$	3200	− ·7
$2(T - J)$	6400	− 1·6
J	299	− ·5
$T - 2J$	2900	− 1·8
$2J$	598	− 1·8
$h + J$	108	− 1·3
$h - 2J$	789	− 1·8
$2h - 2J$	980	− 1·4
$2h - J$	681	− 1·3
$(2h + 2g) - 2J$	204	− ·1
" $- 3J$	93	− ·7
" $- J$	503	− ·9
" $- 4J$	395	− 1·7
$g + h - J$	102	− ·7
$g - 2J$	6·57	− 1·6
$T - M$	1662	− 1·6
$2T - 2M$	3323	− 1·9
$2T - 3M$	1437	− 1·8
$T - 2M$	225	− 1·1
$2T - 4M$	450	− 1·8
$h - T + 2M$	34	− 1·0
$2g + 2h + 3T - 6M$	127	− 1·7
$4h + 5T - 9M$	·80	− ·6
$2g + h + 7T - 4M$	5·46	− 1·4
$l - 26T + 24M$	57	− 1·3
$2\mathrm{D} - l - 20T + 16M$	40	− ·2
$\mathrm{D} - 23T + 20M$	8·65	− ·5
$4\mathrm{D} - l - 2\mathrm{F} - 20(T - M)$	28	− 1·7
$3\mathrm{D} - \mathrm{F} - l - 20T + 18M$	6·04	− ·4
$4\mathrm{D} - 3l - 23T + 25M$	·58	+ 1·4
S	120	computed

SECTION VI.

INEQUALITIES IN LONGITUDE, LATITUDE AND PARALLAX DUE TO THE DIRECT ACTION OF THE PLANETS.

In all cases ϕ denotes the argument of the primary term.

When a primary term consists of two or more terms of the same period but with different arguments, the latter are denoted by $\phi_0 + C$ where C depends on h'', ϖ', ϖ'', and on the coefficients. Each term is set down. The single term, argument ϕ, which is obtained by combining them, is then given. The secondary terms, arguments $\phi+\phi'$, where ϕ' consists of the lunar arguments only, follow. Secondary terms having the same arguments as certain primary terms have not been added to the latter but are given separately in their proper places. (See Addendum.)

Coefficients of Sines in Longitude.

Venus. Short Period Primaries.

$$\phi = i(T - V)$$

i	ϕ	$\phi \pm l$	$\phi \pm (2D - l)$	$\phi \pm 2D$	$\phi \pm 2l$
1	+ 0″·478	+ 0″·076	− 0″·005	+ 0″·006	+ 0″·005
2	+ 195	+ 33		+ 2	+ 2
3	+ 102	+ 19			
4	+ 59	+ 12			
5	+ 35	+ 8			
6	+ 22	+ 5			
7	+ 14	+ 4			
8	+ 9	+ 2			
9	+ 6				
10	+ 5				
11	+ 3				
12	+ 2				

$\phi = l' + i(T - V)$

i	ϕ	$\phi \pm l$
− 3	− 0″·002	
− 2	− 5	
− 1	− 10	
0	− 20	− 0″·003
1	− 65	− 11
2	+ 102	+ 19
3	+ 28	+ 6
4	+ 15	+ 2
5	+ 10	+ 2
6	+ 7	
7	+ 5	
8	+ 3	
9	+ 2	
10	+ 2	

$\phi = l'' + i(T - V)$

i	ϕ	$\phi \pm l$
− 1	+ 0″·001	
0	+ 3	
1	+ 7	
2	+ 20	+ 0″·004
3	− 32	− 7
4	− 9	
5	− 5	
6	− 3	
7	− 2	

$\phi = 2l' + i(T - V)$

i	ϕ	$\phi \pm l$
1	− 0″·002	
2	− 3	
3	− 15	− 0″·002
4	+ 3	
5	+ 1	

$\phi = l' + l'' + i(T - V)$

i	ϕ	$\phi \pm l$
2	+ 0″·002	
3	+ 3	
4	+ 12	+ ·002
5	− 3	

$\phi = iT - (i - 2)V - 2h''$

i	ϕ	$\phi \pm l$
0	− 0″·002	
1	− 3	
2	− 4	
3	− 5	
4	− 7	− 1
5	− 35	− 0″·006
6	+ 7	+ 1
7	+ 3	
8	+ 2	

$\phi = 2l'' + 5(T - V)$

− 0″·002

$\phi = 2D + i(T - V)$

i	ϕ
− 17	− 0″·002
− 16	− 3
− 15	− 3
− 14	− 4
− 13	− 5
− 12	− 7
− 11	− 9
− 10	− 10
− 9	− 12
− 8	− 15
− 7	− 17
− 6	− 20
− 5	− 25
− 4	− 28
− 3	− 32
− 2	− 32
− 1	− 30
0	− 23
1	− 12
2	− 7
3	− 5
4	− 3
5	− 2

$\phi = 2D - l + i(T - V)$

i	ϕ	$\phi + l$	$\phi + l - 2D$	$\phi + 2l$
− 17		+ 0″·002		
− 16		+ 3		
− 15		+ 4		
− 14		+ 6		
− 13		+ 8	− 0″·002	
− 12		+ 10	− 2	
− 11	+ 0″·002	+ 13	− 3	
− 10	+ 3	+ 17	− 3	
− 9	+ 3	+ 22	− 4	
− 8	+ 4	+ 27	− 5	+ 0″·002
− 7	+ 5	+ 32	− 6	+ 2
− 6	+ 6	+ 39	− 8	+ 2
− 5	+ 7	+ 46	− 9	+ 3
− 4	+ 8	+ 53	− 11	+ 3
− 3	+ 10	+ 58	− 12	+ 4
− 2	+ 11	+ 60	− 12	+ 4
− 1	+ 11	+ 57	− 11	+ 4
0	+ 9	+ 46	*	+ 3
1	+ 5	+ 26	− 5	+ 2
2	+ 4	+ 17	− 3	
3	+ 3	+ 12	− 2	
4	+ 2	+ 9	− 2	
5		+ 7	− 1	
6		+ 5	− 1	
7		+ 4		
8		+ 3		
9		+ 2		
10		+ 1		

* The argument vanishes identically and the term disappears.

$\phi = 2D - l' + i(T - V)$

i	ϕ
− 6	− 0″·001
− 5	1
− 4	1
− 3	2
− 2	2
− 1	2
0	1

$\phi = l + i(T - V)$

i	ϕ	$\phi - l$	$\phi - l + 2D$
− 14		+ 0″·001	
− 13		+ 2	
− 12		+ 3	
− 11		+ 3	
− 10		+ 4	
− 9		+ 5	− 0″·001
− 8		+ 6	− 1
− 7		+ 8	− 2
− 6		+ 10	− 2
− 5	+ 0″·002	+ 12	− 2
− 4	+ 3	+ 15	− 3
− 3	+ 3	+ 18	− 4
− 2	+ 4	+ 20	− 4
− 1	+ 5	+ 21	− 4
0	+ 5†	*	− 4
1	+ 3	+ 17	− 3
2	+ 3	+ 13	− 3
3	+ 2	+ 11	− 2
4		+ 9	− 2
5		+ 7	
6		+ 6	
7		+ 5	
8		+ 4	
9		+ 3	
10		+ 2	

* Argument is zero.

† Argument is l and this coefficient therefore produces a slight change in the value of e, which is however too small to affect any term sensibly.

$\phi = l + l' + i(T - V)$

i	ϕ	$\phi - l$
0	insensible	+ 0″·001
1		+ 2
2		+ 2
3		+ 1
4		+ 1
5		+ 1
6		+ 1

$\phi = l - l' - i(T - V)$

i	ϕ	$\phi - l$
− 1	insensible	+ 0″·001
0		+ 1
1		+ 2
2		+ 3
3		+ 3
4		+ 3
5		+ 3
6		+ 2
7		+ 2
8		+ 2
9		+ 2
10		+ 1
11		+ 1

$$\phi = 2D - l - l' - i(T - V)$$

i	ϕ	$\phi + l$	$\phi + l - 2D$
−1		+ 0″·002	
0		+ 4	
1	+ 1	+ 7	− 0″·001
2	+ 1	+ 8	− 2
3	+ 2	+ 9	− 2
4	+ 1	+ 10	− 2
5	+ 1	+ 9	− 2
6	+ 1	+ 9	− 2
7	+ 1	+ 9	− 2
8	+ 1	+ 7	− 1
9		+ 6	− 1
10		+ 6	− 1
11		+ 4	
12		+ 3	
13		+ 3	
14		+ 2	
15		+ 2	

$$\phi = 2D + l + i(T - V)$$

i	ϕ	$\phi - l$
−13	insensible	− 0″·001
−12		− 1
−11		− 2
−10		− 2
− 9		− 3
− 8		− 3
− 7		− 4
− 6		− 4
− 5		− 5
− 4		− 6
− 3		− 6
− 2		− 6
− 1		− 6
0		− 5
1		− 2
2		− 1

$$\phi = 2D - l - l'' - i(T - V)$$

i	ϕ	$\phi + l$
2	insensible	− 0″·002
3		− 2
4		− 3
5		− 3
6		− 3
7		− 3
8		− 3
9		− 2
10		− 2
11		− 2

$$\phi = 2l - 2D + i(T - V)$$

i	ϕ	$\phi - l$	$\phi + l$	$\phi - l + 2D$	$\phi - 2l$	$\phi - 2l + 2D$
−5		− 0″·002				
−4		− 3				
−3		− 4				
−2		− 7				
−1		− 13		+ 0″·003		
0	− 0″·002	− 30		†	− 0″·002	
1	− 4	− 58		+ 12	− 4	
2	− 11	− 140		+ 28	− 10	+ 0″·003
3	− 76*	+ 722	− 0″·007	− 147	+ 49	− 13
4	+ 3	+ 95		− 19	+ 6	− 2
5		+ 46		− 9	+ 3	
6		+ 28		− 6	+ 2	
7		+ 18		− 4		
8		+ 12		− 2		
9		+ 9				
10		+ 6				
11		+ 4				
12		+ 3				
13		+ 2				
14		+ 2				

* $s = -363{\cdot}9$, period $= 9\frac{3}{4}$ years.
† See second note on p. 73.

i	$\phi + 2D$	$\phi - 2D$	$\phi - l - 2D$	$\phi - l + l'$	$\phi - l - l'$	$\phi - l - l' + 2D$	$\phi - 2l + 4D$	$\phi - 3l$
3	− 0″·010	+ 0″·008	+ 0″·007	+ 0″·005	− 0″·004	− 0″·007	− 0″·002	+ 0″·003

$$\phi = 2l - 2D + l' + i(T - V)$$

i	ϕ	$\phi - l$	$\phi - l + 2D$	$\phi - 2l$
1		$-0''\cdot003$		
2		$-\quad 6$		
3		$-\quad 12$	$+0''\cdot002$	
4	$-0''\cdot005$	$-\quad 42$	$+\quad 9$	$-0''\cdot003$
5		$+\quad 34$	$-\quad 7$	$+\quad 2$
6		$+\quad 12$	$-\quad 2$	
7		$+\quad 7$		
8		$+\quad 5$		
9		$+\quad 4$		
10		$+\quad 2$		

$$\phi = 2l - 2D + l'' + i(T - V)$$

i	ϕ	$\phi - l$	$\phi - l + 2D$
3	insensible	$+0''\cdot002$	
4		$+\quad 4$	
5		$+\quad 13$	$-0''\cdot003$
6		$-\quad 10$	$+\quad 2$
7		$-\quad 4$	
8		$-\quad 2$	

$$\phi = 2l - 2D - l' + i(T - V)$$

i	ϕ	$\phi - l$
0	ins.	$-0''\cdot002$
-1		$-\quad 5$

$$\phi = 2l - 2D - l'' + i(T - V)$$

i	ϕ	$\phi - l$
0	ins.	$+0''\cdot004$

$$\phi = 2l - 2D \pm \{(i+2)T - iV\}$$

See long period terms.

$$\phi = h \pm \{(i+2)T - iV\}$$

See long period terms.

Venus. Long Period Primaries.

$\phi_0 = 8T - 3V$, $s = -$ 15[illegible]$''\cdot 0$

period 7·8 years

$\phi_0 - \varpi' - 2h''$	$+0''\cdot 007$
$\phi_0 - \varpi'' - 2h''$	$-\quad 2$
$\phi_0 - 3\varpi'$	$+\quad 1$
$\phi_0 - 2\varpi' - \varpi''$	$-\quad 1$
$\phi_0 + 112° = \phi$	$+0''\cdot 007$

$\phi_0 = 13T - 8V$ $s = -14''\cdot 85$

period 239 years

$\phi_0 - \varpi' - 4h''$	$+0''\cdot 0028$
$\phi_0 - \varpi'' - 4h''$	$-\quad 9$
$\phi_0 - 3\varpi' - 2h''$	$+\quad 31$
$\phi_0 - 2\varpi' - \varpi'' - 2h''$	$-\quad 30$
$\phi_0 - \varpi' - 2\varpi'' - 2h''$	$+\quad 10$
$\phi_0 - 3\varpi'' - 2h''$	$-\quad 1$
$\phi_0 - 39° = \phi$	$+0''\cdot 003$

$\phi_0 = l + 3T - 10V$, $s = +1''\cdot 85$

period 1920 years

$\phi_0 - \varpi' - 6h''$	$+0''\cdot 34$
$\phi_0 - \varpi'' - 6h''$	$-\quad 5$
$\phi_0 - 3\varpi' - 4h''$	$+\quad 7$
$\phi_0 - 2\varpi' - \varpi'' - 4h''$	$-\quad 5$
$\phi_0 - \varpi' - 2\varpi'' - 4h''$	$+\quad 1$
$\phi_0 - 5\varpi' - 2h''$	$+\quad 1$
$\phi_0 - 4\varpi' - \varpi'' - 2h''$	$-\quad 1$
$\phi_0 + 33° = \phi$	$+0''\cdot 35$

$\phi \pm l$	$+0''\cdot 019$
$\phi \pm (l - 2D)$	$+\quad 4$
$\phi \pm 2D$	$+\quad 4$

$\phi_0 = -l - 16T + 18V$, $s = +13''\cdot 01$

period 273 years

$\phi_0 - 2h''$	$-13''\cdot 99$
$\phi_0 - 2\varpi'$	$-\quad 66$
$\phi_0 - \varpi' - \varpi''$	$+\quad 71$
$\phi_0 - 2\varpi''$	$-\quad 20$
$\phi_0 - \varpi' + \varpi'' - 2h''$	$-\quad 8$
$\phi_0 + \varpi' - \varpi'' - 2h''$	$-\quad 27$
$\phi_0 + 2\varpi' - 4h''$	$-\quad 6$
$\phi_0 + \varpi' + \varpi'' - 4h''$	$+\quad 5$
$\phi_0 + 2\varpi'' - 4h''$	$-\quad 1$
$\phi_0 - 151° \, 00' = \phi$	$-14''\cdot 55$

$\phi + l$	$-0''\cdot 766$
$\phi - l$	$-\quad 815$
$\phi + 2D - l$	$-\quad 159$
$\phi - 2D + l$	$-\quad 168$
$\phi \pm (2D + l)$	$-\quad 20$
$\phi + 2D$	$-\quad 166$
$\phi - 2D$	$-\quad 168$
$\phi + 2l$	$-\quad 52$
$\phi - 2l$	$-\quad 55$
$\phi \pm (2D - l')$	$-\quad 11$
$\phi \pm D$	$+\quad 5$
$\phi \pm (l + l')$	$+\quad 4$

$\phi \pm (2D - l - l')$	$-0''\cdot 008$
$\phi + l - l'$	$-\quad 5$
$\phi - l + l'$	$-\quad 6$
$\phi \pm 2F$	$+\quad 29$
$\phi \pm 3l$	$-\quad 4$
$\phi \pm (4D - 2l)$	$-\quad 2$
$\phi \pm (4D - l)$	$-\quad 4$
$\phi \pm (2F + l)$	$+\quad 5$
$\phi \pm (2F - l)$	$+\quad 2$

$\phi_0 = l + 29T - 26V,\ s = -27''\cdot 85$

period 127 years

$\phi_0 - \varpi' - 2h''$	$+0''\cdot 096$
$\phi_0 - \varpi'' - 2h''$	$-\ \ 25$
$\phi_0 - 3\varpi'$	$+\ \ 79$
$\phi_0 - 2\varpi' - \varpi''$	$-\ \ 71$
$\phi_0 - \varpi' - 2\varpi''$	$+\ \ 22$
$\phi_0 - 3\varpi''$	$-\ \ 2$
$\phi_0 - 2\varpi' + \varpi'' - 2h''$	$+\ \ 3$
$\phi_0 + \varpi' - 2\varpi'' - 2h''$	$-\ \ 1$
$\phi_0 - 4\varpi' + \varpi''$	$-\ \ 1$
$\phi_0 + 112° = \phi$	$+0''\cdot 108$
$\phi \pm l$	$+0''\cdot 006$

$\phi = l + 21(T - V),\ s = 425''\cdot 0$

period 8·4 years

ϕ	$+\ 0''\cdot 030$
$\phi - l$	$+\ 0''\cdot 005$
$\phi + l$	$+\ 0''\cdot 002$

$\phi = l + 22(T - V),\ s = -1795''$

period 2·0 years

ϕ	$+0''\cdot 002$

$\phi_0 = l + 24T - 23V,\ s = -465''\cdot 8$

period 7·6 years

$\phi_0 - \varpi'$	$+0''\cdot 009$
$\phi_0 - \varpi''$	$-\ \ 3$
$\phi_0 - 88° = \phi$	$+0''\cdot 006$

$\phi = 2D - l + 18(T - V),\ s = 788''\cdot 8$

period 3·5 years

ϕ	$+0''\cdot 011$
$\phi + l$	$+0''\cdot 003$

$\phi = 2D - l + 19(T - V),\ s = -1430''\cdot 7$

period 2·5 years

ϕ	$+0''\cdot 002$

$\phi_0 = l - 2D - 8T + 12V,\ s = +87''\cdot 07$

period 41 years

$\phi_0 - 4h''$	$-0''\cdot 032$
$\phi_0 - 2\varpi' - 2h''$	$-\ \ 2$
$\phi_0 - \varpi' - \varpi'' - 2h''$	$+\ \ 1$
$\phi_0 - 123° = \phi$	$+0''\cdot 033$

$\phi = l - 2D - 13T + 15V - 2h'',\ s = -351''$

period 10 years

ϕ	$-0''\cdot 025$
$\phi - l$	$-0''\cdot 004$

$\phi_0 = l - 2D - 16T + 17V,\ s = 539''\cdot 9$

period 6·6 years

$\phi_0 - \varpi'$	$+0''\cdot 004$
$\phi_0 - \varpi''$	$-\ \ 2$
$\phi_0 - 73° = \phi$	$+0''\cdot 003$

$\phi_0 = 2D - l + 21T - 20V,\ s = -101''\cdot 9$

period 35 years

$\phi_0 - \varpi'$	$+0''\cdot 171$
$\phi_0 - \varpi''$	$-\ \ 52$
$\phi_0 + \varpi' - 2h''$	$-\ \ 3$
$\phi_0 + \varpi'' - 2h''$	$+\ \ 1$
$\phi_0 - 87°\ 00' = \phi$	$+0''\cdot 126$
$\phi + l$	$+0''\cdot 010$
$\phi - l$	$+0''\cdot 007$

$\phi_0 = 2l - 2D + 8T - 6V$, $s = 74''{\cdot}06$

period 48 years

$\phi_0 - 2h''$	$-$	$0''{\cdot}036$
$\phi_0 - 2\varpi'$	$-$	56
$\phi_0 - \varpi' - \varpi''$	$+$	35
$\phi_0 - 2\varpi''$	$-$	5
$\phi_0 + 17^\circ\ 30' = \phi$	$+$	$0''{\cdot}065$
$\phi - 2D$	$+$	$0''{\cdot}002$
$\phi - l$	$+$	86
$\phi + l$	$+$	5
$\phi + 2D - l$	$-$	16
$\phi - 2l$	$+$	6
$\phi - 2l + 2D$	$-$	2

$\phi = 2l - 2D + 7T - 5V + 15^\circ$, $s = 2296''$

period 1·5 years

ϕ	insensible
$\phi - l$	$+0''{\cdot}004$

$\phi = 2l - 2D + 9T - 7V + 20^\circ$, $s = -2145''$

period 1·7 years

ϕ	insensible
$\phi - l$	$-0''{\cdot}003$

$\phi = 2F - 2D + 3(T - V)$, $s = 819''{\cdot}5$

period 4·3 years

ϕ	$-0''{\cdot}002$
$\phi - 2F$	$+0''{\cdot}002$

$\phi_0 = 2F - 2D + 6T - 5V$, $s = -71''{\cdot}27$

period 50 years

$\phi_0 - \varpi'$	$+0''{\cdot}071$
$\phi_0 - \varpi''$	$-$ 21
$\phi_0 - 90^\circ = \phi$	$+0''{\cdot}054$
$\phi \pm l$	$+0''{\cdot}003$
$\phi - 2F$	$-0''{\cdot}003$

$\phi = 3l - 2D + 24(T - V)$, $s = 61''{\cdot}07$

period 58 years

ϕ	$+0''{\cdot}010$

$\phi = 4D - l - 2F + 15(T - V)$, $s = -30''{\cdot}7$

period 116 years

ϕ	$-0''{\cdot}002$

ϕ	s	period	coef. primary	secondary
$h - 4T + 2V + h''$	$-2848''$	$1\frac{1}{4}$ years	$+0''{\cdot}005$	
$h - 5T + 3V + h''$	$-\ 628''{\cdot}7$	5·2 years	$+0''{\cdot}016$	$\phi - 2F$ $+0''{\cdot}003$
$h - 6T + 4V + h''$	$+\ 1590{\cdot}8$	2·2 years	$-0''{\cdot}009$	
$h - 7T + 5V + h''$	$+\ 3810{\cdot}$	0·9 years	$-0''{\cdot}003$	

$\phi_0 = l + 2g + h + 19T - 20V$, $s = 88''{\cdot}85$

period 40 years

$\phi_0 + \varpi' - h''$	$-$	$0''{\cdot}003$
$\phi_0 + \varpi'' - h''$	$+$	2
$\phi_0 + 166^\circ = \phi$	$+$	$0''{\cdot}002$

$\phi = l + h + 16T - 18V + h''$, $s = -203''{\cdot}8$

period 17 years

ϕ	$+0''{\cdot}008$

$\phi = l + g + 24T - 23V - h''$, $s = 125''\cdot 9$

period 28 years

ϕ	$+0''\cdot 003$

$\phi = l + 4g + 3h + 24T - 22V - 3h''$, $s = -0''\cdot 07$*

period 50000 years

ϕ	$+0''\cdot 02$

$\phi_0 = -D - 12T + 15V$, $s = 50''\cdot 0$

period 71 years

$\phi_0 - \varpi' - 2h''$	$-0''\cdot 008$
$\phi_0 - \varpi'' - 2h''$	$-$ 5
$\phi_0 - 82^\circ = \phi$	$+0''\cdot 013$

$\phi = D + 20(T - V)$, $s = -502''\cdot 8$

period 7·1 years

ϕ	$+0''\cdot 002$

$\phi_0 = D + 25T - 23V$, $s = -64''\cdot 9$

period 54 years

$\phi_0 - 2h''$	$+0''\cdot 006$
$\phi_0 - 2\varpi'$	$+$ 12
$\phi_0 - \varpi' - \varpi''$	$-$ 6
$\phi_0 - 2\varpi''$	$+$ 1
$\phi_0 + 190^\circ = \phi$	$+0''\cdot 013$

$\phi_0 = l - D + 4T - 3V$, $s = 37''\cdot 03$

period 94 years

$\phi_0 - \varpi'$	$-0''\cdot 059$	from term of R factored by $\frac{a}{a'}$
$\phi_0 - \varpi''$	$+$ 18	
$\phi_0 + \varpi' - 2h''$	$-$ 1	
$\phi_0 + 93^\circ$	$+0''\cdot 044$	

$\phi_0 - \varpi'$	$+0''\cdot 005$	from first part of R
$\phi_0 - \varpi''$	$-0''\cdot 002$	
$\phi_0 - 88^\circ$	$+0''\cdot 004$	

$\phi_0 + 93^\circ = \phi$	$+0''\cdot 040$
$\phi - l$	$+0''\cdot 029$
$\phi + l$	$+0''\cdot 002$
$\phi + 2D - l$	$-0''\cdot 005$
$\phi - 2l$	$+0''\cdot 002$

$\phi_0 = 3D - 2F + 19T - 18V$, $s = 6''\cdot 33$

period 560 years

$\phi_0 - \varpi'$	$+0''\cdot 003$
$\phi_0 - \varpi''$	$-0''\cdot 001$
$\phi_0 - 88^\circ = \phi$	$+0''\cdot 002$

$\phi = h - h''$

Venus	$-0''\cdot 0152$
Mercury	$-$ 1
Jupiter	$+$ 24
Mars	$+$ 3
ϕ	$-0''\cdot 013$

* This period is approximate only. But the coefficient is insensible to observation whatever be the period.

Jupiter. Short Period Primaries.

$\phi = i\,(T - J)$

i	ϕ	$\phi \pm l$
1	− 0″·069	− 0″·008
2	− 0″·001	− 0″·005

$\phi = J - i\,(T - J) + 174°$

i	ϕ	$\phi \pm l$
0	See long period primaries	
1	− 0″·011	− 0″·002
2	− 0″·002	

$\phi = 2D + i\,(T - J)$

i	ϕ
1	− 0″·002
2	− 22
3	− 10
4	− 3

$\phi = l + i\,(T - J)$

i	ϕ	$\phi - l$
− 1		+ 0″·003
0		*
1		+ 0″·003
2	+ 0″·002	+ 7
3	+ 1	+ 3

* The argument is zero.

$\phi = l - 2D + i\,(T - J)$

i	ϕ	$\phi - l$	$\phi - l + 2D$
0		− 0″·002	
− 1		− 4	
− 2	− 0″·007	− 39	+ 0″·008
− 3	− 3	− 17	+ 3
− 4		− 5	

$\phi = 2l - 2D + i\,(T - J)$

i	ϕ	$\phi - l$	$\phi - l + 2D$	$\phi - 2l$
0	insensible	− 0″·003		
− 1	insensible	− 7	+ 0″·001	
− 2	See long period primaries			
− 3	insensible	+ 23	− 0″·005	+ 0″·002
− 4	insensible	+ 4		

$\phi = 2l - 2D + l'' + i\,(T - J)$

i	ϕ	$\phi - l$
− 3	insensible	+ 0″·006
− 2	See long period primaries	

$\phi = 2D + l + i\,(J - T)$

i	ϕ	$\phi - l$
− 2	− 0″·001	− 0″·005
− 3		− 0″·002

Jupiter. Long Period Primaries.

$\phi_0 = J,\ s = 299''\cdot 13$

period 11·8 years

$\phi_0 - \varpi''$		$-0''\cdot 208$
$\phi_0 - \varpi'$		$+0''\cdot 021$
$\phi_0 + 173^\circ\ 50' = \phi$		$+0''\cdot 209$
$\phi \pm l$		$+0''\cdot 021$
$\phi \pm 2D$		$+0''\cdot 002$

$\phi_0 = 2J,\ s = 598''\cdot 3$

period 5·9 years

$\phi_0 - 2\varpi''$	$-0''\cdot 008$
$\phi_0 - \varpi' - \varpi''$	$+0''\cdot 001$
$\phi_0 - 2h''$	$-0''\cdot 001$
$\phi_0 + 162^\circ = \phi$	$+0''\cdot 009$

$\phi = 2l - 2D - 2(T - J),\ s = -203''\cdot 58$

period 15·5 years

ϕ		$-0''\cdot 187$
$\phi - l$		$+0''\cdot 811$
$\phi + l$	$-$	7
$\phi - l + 2D$	$-$	·170
$\phi - l - 2D$	$+$	7
$\phi - 2l$	$+$	56
$\phi - 2l + 2D$	$-$	15
$\phi - l - l'$	$-$	4
$\phi - l - l' + 2D$	$-$	8
$\phi - l + l'$	$+$	5
$\phi - 3l$	$+$	4
$\phi + 2D$	$-$	13
$\phi - 2D$	$+$	9
$\phi - 2l + 4D$	$-$	2

$\phi_0 = 2l - 2D + 3J - 2T,\ s = 95''\cdot 55$

period 37 years

$\phi_0 - \varpi''$		$-0''\cdot 190$
$\phi_0 - \varpi'$		$+0''\cdot 017$
$\phi_0 + 173^\circ\ 16' = \phi$		$+0''\cdot 190$
$\phi - l$		$+0''\cdot 306$
$\phi + l$	$+$	5
$\phi - l + 2D$	$-$	58
$\phi - l - 2D$	$+$	3
$\phi - 2l$	$+$	20
$\phi - 2D$	$+$	6
$\phi + 2D$	$-$	2
$\phi - 2l + 2D$	$-$	6
$\phi - l - l' + 2D$	$-$	3
$\phi - l + l'$	$+$	2
$\phi - l - l' + 2D$	$-$	3
$\phi - l + l'$	$+$	2

$\phi = 2l - 2D + J - 2T + \varpi'',\ s = -502''\cdot 6$

period 7 years

ϕ	insensible
$\phi - l$	$-0''\cdot 007$

$\phi = 2l - 2D + 4J - 2T - 2\varpi'',\ s = 394''\cdot 7$

period 9 years

ϕ	$-0''\cdot 002$
$\phi - l$	$-0''\cdot 009$
$\phi - l + 2D$	$+0''\cdot 002$

$\phi = 2D - 2F + 2(T - J),\ s = -979''\cdot 8$

period 3·6 years

ϕ	$+0''\cdot 002$

$\phi = h + J - h'',\ s = 108''\cdot 36$

period 33 years

ϕ	$-0''\cdot 004$

Mars. Short Period Primaries.

$\phi = i(T - M)$

i	ϕ	[illegible]
1	$-0''{\cdot}019$	$-0''{\cdot}002$
2	$-\quad 8$	
3	$-\quad 3$	
4	$-\quad 2$	

$\phi_0 = M + i(M - T)$

i	[illegible]	[illegible]
-1	$-0''{\cdot}002$	
0	$-\quad 5$	
1	$-\quad 55$	$+0''{\cdot}008$ See long period
2	$+\quad 9$	$-\quad 1$ primaries
3	$+\quad 4$	

$\phi_0 = 2M + i(M - T)$

i	$\phi_0 - 2\varpi''$	$\phi_0 - \varpi' - \varpi''$	$\phi = \phi_0 + 63°$
1	$-0''{\cdot}002$		$-0''{\cdot}002$
2	$-\quad 10$	$+0''{\cdot}003$	$-\quad 12$
3	$+\quad 4$	$-\quad 1$	$+\quad 5$
4	$+\quad 2$		$+\quad 2$
5	$+\quad 1$		$+\quad 1$

$\phi = 3M + 3(M - T) - 3\varpi''$

i	ϕ
3	$-0''{\cdot}002$
4	$+\quad 1$

$\phi = 2l - 2D + i(M - T)$

i	ϕ	$\phi - l$	$\phi - l + 2D$
2		$-0''{\cdot}002$	
3		$-\quad 5$	
4	$-0''{\cdot}002$	$+\quad 17$	$-0''{\cdot}003$
5		$+\quad 3$	
6		$+\quad 1$	

$\phi_0 = 2l - 2D + i(T - M) + M$

i	$\phi_0 - \varpi''$	$\phi_0 - \varpi'$	$\phi_0 + 211° - l$
-4			$-0''{\cdot}003$
-5	$-0''{\cdot}014$	$+0''{\cdot}002$	See long period primaries
-6			$+0''{\cdot}003$

Mars. Long Period Primaries.

$\phi_0 = 2M - T,\ s = 224''\cdot8$

period 16 years

$\phi_0 - \varpi''$	$-0''\cdot055$
$\phi_0 - \varpi'$	$+\quad 8$
$\phi_0 + 213^\circ = \phi$	$+0''\cdot060$
$\phi \pm l$	$+0''\cdot007$

$\phi_0 = 2l - 2D + 6M - 5T,\ s = -127''\cdot31$

period 28 years

$\phi_0 - \varpi''$	$-0''\cdot014$
$\phi_0 - \varpi'$	$+\quad 2$
$\phi_0 + 211^\circ = \phi$	$+0''\cdot015$
$\phi - l$	$-0''\cdot038$
$\phi - l + 2D$	$+\quad 8$
$\phi - 2l$	$-\quad 3$

$\phi_0 = 2l - 2D + 8M - 6T,\ s = +97''\cdot56$

period 38 years

$\phi_0 - 2\varpi''$	$-0''\cdot015$
$\phi_0 - \varpi' - \varpi''$	$+\quad 5$
$\phi_0 - 2\varpi'$	$-\quad 1$
$\phi_0 + 63^\circ = \phi$	$-0''\cdot019$
$\phi - l$	$-0''\cdot030$
$\phi - l + 2D$	$+\quad 6$
$\phi - 2l$	$-\quad 2$
$\phi + l$	$-\quad 1$

$\phi_0 = 4D - 3l + 25M - 23T,\ s = -0''\cdot58$

period 6000 years

$\phi_0 - 2\varpi''$	$+0''\cdot03$
$\phi_0 - \varpi' - \varpi''$	$-\quad 1$
$\phi_0 - 113^\circ = \phi$	$-0''\cdot04$
$\phi \pm l$	$-0''\cdot002$

$\phi_0 = h + 2M - T,\ s = 34''\cdot07$

period 104 years

$\phi_0 - \varpi'' - h''$	$-0''\cdot015$
$\phi_0 - \varpi' - h''$	$+\quad 2$
$\phi_0 + 165^\circ = \phi$	$+0''\cdot017$

Mercury. No Short Period Primaries.

Long Period Primaries.

$\phi_0 = l - 2D - T + 3Q,\ s = -90''\cdot36$

period 39 years

$\phi_0 - 2\varpi''$	$-0''\cdot047$
$\phi_0 - \varpi' - \varpi''$	$+\quad 3$
$\phi_0 - 2h''$	$-\quad 40$
$\phi_0 + 75^\circ = \phi$	$+0''\cdot075$
$\phi - l$	$+0''\cdot006$
$\phi + l$	$+0''\cdot004$

$\phi_0 = 2D - l + 5T - 4Q,\ s = -449''\cdot3$

period 7·9 years

$\phi_0 - \varpi''$	$-0''\cdot004$
$\phi_0 - \varpi'$	$+\quad 1$
$\phi_0 + 113^\circ = \phi$	$+0''\cdot003$

$\phi_0 = -2F + l - 3T + 4Q,\ s = 67''\cdot75$

period 52 years

$\phi_0 - \varpi''$	$-0''\cdot004$
$\phi_0 - \varpi'$	$+\quad 1$
$\phi_0 + 113^\circ = \phi$	$+0''\cdot003$

$\phi = l + 2g + 3h - T - 3Q + h'',\ s = -100''\cdot41$

period 35 years

ϕ	$+0''\cdot002$

Saturn.

Short Period Primaries.

$\phi = 2D - l - 2(S - T)$

ϕ	insensible
$\phi + l$	$+0''\cdot003$

$\phi = 2D - 2(S - T)$

ϕ	$-0''\cdot001$

$\phi = 2l - 2D - 2(T - S),\ s = -560''\cdot9$

period 6·3 years

ϕ	insensible
$\phi - l$	$+0''\cdot014$
$\phi - l + 2D$	$-\quad 3$

Long Period Primaries.

$\phi_0 = S,\ s = 120''\cdot45$

period 30 years

$\phi_0 - \varpi''$	$-0''\cdot026$
$\phi_0 - \varpi'$	$+\quad 2$
$\phi_0 + 90^\circ = \phi$	$+0''\cdot024$
$\phi \pm l$	$+0''\cdot003$

$\phi = 2l - 2D + 3S - 2T - \varpi'',\ s = -440''\cdot5$

period 8 years

ϕ	insensible
$\phi - l$	$+0''\cdot004$

Coefficients of Sines in Latitude.

Venus. Short Period Primaries.

$\phi = i(T - V)$

i	$\phi \pm F$	$\phi \pm F \pm l$
1	$+0''\cdot005$	$+0''\cdot002$
2	$+\quad 3$	

$\phi = 2F - 2D + i(T - V)$

i	$\phi - F$
0	$-0''\cdot003$
1	$-\quad 6$
2	$-\quad 11$
3	See long period primaries
4	$+\quad 25$
5	$+\quad 9$
6	$+\quad 5$
7	$+\quad 3$
8	$+\quad 2$

$\phi = 2D - l + i(T - V)$

i	$\phi \pm F + l$
-4	$+0''\cdot002$
-3	$+\quad 3$
-2	$+\quad 3$
-1	$+\quad 3$
0	$+\quad 2$

$\phi = 2l - 2D + i(T - V)$

i	$\phi \pm F - l$
1	$-0''\cdot002$
2	$-\quad 6$
3	See long period primaries
4	$+\quad 4$
5	$+\quad 2$

$\phi = h + h'' - iT + (i-2)V$

i	$\phi + F$
−1	+0″·003
0	+ 4
1	+ 6
2	+ 8
3	+ 10
4	+ 18
5	See long period primaries
6	− 25
7	− 9
8	− 5
9	− 3
10	− 2

$\phi = 2F - 2D + T + i(T - V) - 90°$

i	$\phi - F$
4	+0″·002
5	See long period primaries
6	−0″·002

$\phi = h - h'' + i(T - V)$

i	$\phi + F$
−7	+0″·002
−6	+ 3
−5	+ 5
−4	+ 6
−3	+ 9
−2	+ 14
−1	+ 27
0	See long period primaries
1	− 15
2	− 6
3	− 3

Long Period Primaries.

$\phi = 2l - 2D + 3(T - V)$, $s = -363''\cdot 9$

period $9\frac{3}{4}$ years

$\phi \pm F$	−0″·003
$\phi \pm F - l$	+ 32
$\phi + F - l + 2D$	− 6
$\phi - F - l + 2D$	− 5
$\phi - F - 2l$	+ 4

$\phi = 2F - 2D + 6T - 5V - 90°$, $s = -71''\cdot 27$

period 50 years

$\phi - F$	+0″·068
$\phi - F + 2D$	− 2
$\phi - F - l$	+ 4
$\phi - F + l$	− 4

$\phi = h - h''$, $s = -190''\cdot 8$

period 18·6 years

$\phi + F$	−0″·241
$\phi + F - 2D$	+ 8
$\phi + F + l$	− 13
$\phi + F - l$	+ 13

$\phi = 2l - 2D + 8T - 6V + 17°30'$, $s = 74''\cdot 06$

period 48 years

$\phi \pm F$	+0″·003
$\phi \pm F - l$	+ 4

$\phi = 2F - 2D + 3(T - V)$, $s = 819''\cdot 5$

period $4\frac{1}{3}$ years

$\phi - F$	−0″·044
$\phi - F - l$	− 2
$\phi - F + l$	+ 2

$\phi = h + h'' - 5T + 3V$, $s = -628''\cdot 72$

period 5·6 years

$\phi + F$	+0″·072
$\phi + F - 2D$	− 2
$\phi + F + l$	+ 4
$\phi + F - l$	− 4

$\phi = l + 3T - 10V + 33°,\ s = +1''\cdot 85$

period 1920 years

$\phi + F$	$+0''\cdot016$

$\phi = l + 29T - 26V + 112°,\ s = -27''\cdot85$

period 127 years

$\phi \pm F$	$+0''\cdot005$

$\phi = 2D - l + 21T - 20V - 87°0',\ s = -101''\cdot92$

period 35 years

$\phi \pm F$	$+0''\cdot006$

$\phi = -l - 16T + 18V - 150°59',\ s = 13''\cdot01$

period 273 years

$\phi \pm F$		$-0''\cdot650$
$\phi + (F - 2D)$	+	22
$\phi \pm (F + 3D)$	−	12
$\phi + F + l$	−	79
$\phi - F - l$	−	56
$\phi + F - l$	−	7
$\phi - F + l$	+	14
$\phi + F - l + 2D$	−	13
$\phi - F + l - 2D$	−	17
$\phi + F + 2l$	−	8
$\phi + F - 2l$	−	6
$\phi \pm (F - l - 2D)$	−	3

Jupiter. No Short Period Primaries.

Long Period Primaries.

$\phi = J - \varpi'',\ s = 299''\cdot1$

period 12 years

$\phi \pm F$	$+0''\cdot002$

$\phi = 2l - 2D + 3J - 2T + 173°16',\ s = 95''\cdot55$

period 37 years

$\phi \pm F$		$+0''\cdot009$
$\phi - l \pm F$	+	14
$\phi - l - F + 2D$	−	2
$\phi - l + F + 2D$	−	3

$\phi = 2l - 2D - 2(T - J),\ s = -203''\cdot58$

period 15·5 years

$\phi \pm F$		$-0''\cdot008$
$\phi - F - l$	+	35
$\phi + F - l$	+	36
$\phi - F - l + 2D$	−	6
$\phi + F - l + 2D$	−	7
$\phi - F - 2l$	+	4
$\phi + F - 2l$	+	2

$\phi = h - h'' + J - \varpi'',\ s = 108''\cdot36$

period 33 years

$\phi + F$	$-0''\cdot005$

$\phi = h - h'',\ s = -190''\cdot8$

period 19 years

$\phi + F$	$+0''\cdot038$

$\phi = 2F - 2D - 2(T - J),\ s = 979''\cdot8$

period 3·6 years

$\phi - F$	$-0''\cdot026$

$\phi = h - h'' + T - J,\ s = 3058''$

period 1·2 years

$\phi + F$	$-0''\cdot002$

$\phi = 2F - 2D - 3(T - J),\ s = -2269''$

period 1·6 years

$\phi - F$	$+0''\cdot005$

Mars.

$\phi = h + 2M - T + 165°$, $s = 34''\cdot07$		$\phi = h - h''$, $s = -190''\cdot8$	
period 104 years		period 19 years	
$\phi + F$	$-0''\cdot010$	$\phi + F$	$+0''\cdot004$

Mercury and Saturn. None.

Coefficients of Cosines in Parallax.

$\phi = 2l - 2D + 3(T - V)$	$\phi - l$	$+0''\cdot006$
$\phi = l + 16T - 18V + 151°$	$\phi \pm l$	$+0''\cdot007$
$\phi = 2l - 2D - 2(T - J)$	$\phi - l$	$+0''\cdot007$
$\phi = 2l - 2D + 3J - 2T + 173°$	$\phi - l$	$+0''\cdot003$

Additions to Annual Mean Motions.

Perigee	$+2''\cdot69$
Node	$-1''\cdot42$

ADDENDUM. FINAL VALUES.

The primary coefficients in longitude having short periods can be combined with the secondary coefficients having the same arguments in many cases. These combinations have been made in the following tables which therefore represent the final coefficients due to the direct action of the planets, with the following exceptions :—

Venus. All long period primaries together with their secondaries except those arising from the arguments

$$2l - 2\mathrm{D} + 2T + i(T - V) + a, \quad 2F - 2\mathrm{D} + 3(T - V), \quad 2F - 2\mathrm{D} + 6T - 5V - 90^\circ,$$

which have been included.

Mars. The long period primaries with arguments

$$4\mathrm{D} - 3l + 25M - 23T - 113^\circ, \quad h + 2M - T + 165^\circ.$$

Saturn and Mercury. All terms.

These terms, omitted here, can be taken immediately from the tables in Sect. VI.

Final coefficients of sines of $i(T - V)$ + *arguments at top of each column.*

i	0
1	+ 0″·480
2	+ ·200
3	+ 92
4	+ 60
5	+ 38
6	+ 25
7	+ 17
8	+ 12
9	+ 8
10	+ 6
11	+ 4
12	+ 1

$T - 87^\circ\cdot8$	
− 3	− 0″·001
− 2	− 4
− 1	− 8
0	− 15
1	− 47
2	+ 76
3	+ 21
4	+ 12
5	+ 7
6	+ 6
7	+ 4
8	+ 1

$2T + 26^\circ\cdot6$	
− 4	+ 0″·001
− 3	+ 2
− 2	+ 2
− 1	+ 3
0	+ 4
1	+ 6
2	+ 8
3	+ 37
4	− 8
5	− 3
6	− 4
7	− 1

2D	
− 15	+ 0″·001
− 14	+ 2
− 13	+ 2
− 12	+ 2
− 11	+ 3
− 10	+ 5
− 9	+ 6
− 8	+ 8
− 7	+ 8
− 6	+ 11
− 5	+ 11
− 4	+ 10
− 3	− 36
− 2	+ 26
− 1	+ 15
0	+ 16
1	+ 15
2	+ 8
3	+ 4
4	+ 4
5	+ 4
6	+ 3
7	+ 3
8	+ 3
9	+ 2
10	+ 1

$2\mathrm{D} - T + 78^\circ$	
− 15	+ 0″·001
− 14	+ 1
− 13	+ 2
− 12	+ 2
− 11	+ 3
− 10	+ 4
− 9	+ 4
− 8	+ 4
− 7	+ 5
− 6	+ 5
− 5	+ 7
− 4	+ 7
− 3	+ 4
− 2	+ 4
− 1	+ 3
0	+ 2
1	+ 1

$2\mathrm{D} - 2T - 18^\circ$	
− 6	− 0″·006

l

−8	−	0″·002
−7	−	4
−6	−	5
−5	−	6
−4	−	9
−3	−	16
−2	−	29
−1	−	68
0	[+	11]
1	+	91
2	+	64
3	−	127
4	−	7
5	−	1

$l + T - 88°$

0	−	0″·002
1	−	8
2	+	13
3	+	6
4	+	8
5	−	4
6	−	2

$l - T + 88°$

−5	−	0″·001
−4	−	2
−3	−	5
−2	−	13
−1	+	8
0	+	2

$l - T + 100°$

3	−	0″·007

$\pm l + 2T + 27°$

2	+	0″·001
3	+	6
4	−	1

$l + 2T + 17°\cdot5$

6	−	0″·016

$2D - l$

−13	−	0″·001
−12	−	2
−11	−	2
−10	−	3
−9	−	6
−8	−	8
−7	−	13
−6	−	22
−5	−	39
−4	−	87
−3	−	716
−2	+	152
−1	+	74
0	+	39
1	+	13
2	+	10
3	+	7
4	+	5
5	+	3
6	+	2

$2D - l + T - 50°$

−1	+	0″·002

$2D - l - T + 89°$

−10	−	0″·001
−9	−	2
−8	−	3
−7	−	5
−6	−	8
−5	−	25
−4	+	33
−3	+	10
−2	+	5
−1	+	3
0	+	1

$2D - l - 2T - 18°\cdot2$

−5	−	0″·004
−6	−	83
−7	+	3

$2D + l$

−9	+	0″·001
−8	+	2
−7	+	2
−6	+	2
−5	+	3
−4	+	1
−3	+	4
−2	+	4
−1	+	3
0	+	2
1	+	1

$2l$

−3	−	0″·001
−2	−	2
−1	−	5
0		
1	+	5
2	+	2
3	−	9

$2l - 2D$

0	−	0″·002
1	−	4
2	−	11
3	−	76
4	+	3

$2l - 2D + T - 88°$

4	−	0″·004

$2l - 2D + 2T + 17°\cdot5$

6	+	0″·065

$3l - 2D$

3	−	0″·003

$2l - 4D$

3	+	0″·008

$-4D$

3	+	0″·007

$2F - 2D$

3	−	0″·002

$2F - 2D + T - 90°$

5	+	0″·054

$2F - 2D \pm l + T - 90°$

5	+	0″·003

Final coefficients of sines of $i(M - T)$ + arguments at top of each column.

i	0
1	+ 0″·019
2	+ 8
3	+ 3
4	+ 2
5	+ 1

$M + 212°·7$	
− 1	+ 0″·002
0	+ 5
1	+ 60
2	− 10
3	− 4
4	− 2
5	− 1

$2M + 63°$	
1	− 0″·002
2	− 12
3	+ 5
4	+ 2
5	+ 1

$3M + 96°$	
3	− 0″·002
4	+ 1

$2D - M + 149°$	
− 5	+ 0″·003

$2D - 2M + 117°$	
− 6	− 0″·002

$\pm l$	
− 1	− 0″·002

$l + M + 212°$	
1	+ 0″·007
2	− 1
5	+ 8

$l - M + 147°$	
− 2	+ 0″·001
− 1	− 7

$l + 2M + 63°$	
2	− 0″·001
6	+ 6

$l - 2M - 63°$	
− 2	+ 0″·001

$2D - l$	
− 6	− 0″·001
− 5	− 3
− 4	− 17
− 3	+ 5
− 2	+ 2

$2D - l - M + 149°$	
− 7	+ 0″·001
− 6	+ 3
− 5	+ 38
− 4	− 3
− 3	− 1

$2D - l - 2M + 117°$	
− 6	− 0″·030

$2l - 2D$	
4	− 0″·002

$2l - 2D + M + 211°$	
5	+ 0″·015

$2l - 2D + 2M + 63°$	
6	− 0″·019

$3l - 2D + 2M + 63$	
6	− 0″·001

Final coefficients of sines of $i(J-T)+$ arguments at top of each column.

i	0
1	+0″·069
2	− 13

i	$J-6°·2$
0	−0″·209
1	+ 11
2	+ 8

i	$2J-18°$
0	−0″·009

i	$2D$
−5	+0″·001
−4	+ 2
−3	+ 3
−2	− 45
−1	+ 2
0	+ 2
1	
2	− 2

i	$2D+J-6°$
0	−0″·002

i	$2D-J+7°$
−2	+0″·020
0	+ 2

i	l
−3	+0″·001
−2	+ 4
−1	+ 8
0	
1	− 8
2	− 171

i	$l+15°$
3	+0″·002

i	$l+J-6°·7$
0	−0″·021
1	+ 2
2	+ 58
3	− 1

i	$l-J+6°$
−1	−0″·002
0	+ 21

i	$l-J+106°$
3	−0″·008

i	$l+2J-18°$
2	+0″·002

i	$h+J-h''$
0	−0″·004

i	$2D-l$
−4	−0″·004
−3	− 20
−2	− 804
−1	+ 7
0	+ 3

i	$2D-l+J-100°$
−3	+0″·004

i	$2D-l+J-7°$
−2	+0″·007

i	$2D-l-J+6°·73$
−4	−0″·001
−3	− 6
−2	+ 306

i	$2D-l-J-80°$
−1	+0″·005

i	$2D-l-2J+18°$
−2	+0″·009

i	$2D-l-2J+107°$
−1	+0″·002

i	$2D+l$
−2	−0″·003
2	− 1

i	$2D+l-J+7°$
−2	+0″·001

i	$4D-l$
−2	−0″·007

i	$4D-l-J+7°$
−2	+0″·003

i	$2l$
2	−0″·013

i	$2l+J-7°$
2	+0″·002

i	$2l-2D$
2	−0″·187

i	$2l-2D+J-6°·7$
2	−0″·190

i	$2l-2D+2J-18°$
2	−0″·002

i	$2l-4D$
2	+0″·009

i	$2l-4D+J-7°$
2	−0″·006

i	$3l-2D$
2	−0″·007

i	$3l-2D+J-7°$
2	−0″·005

i	$2F-2D$
2	−0″·002

i	$2F+l-2D$
2	−0″·002

i	$-2F+l-2D$
2	+0″·001

ERRATA.

Page 10, line 7, „ 13, „ 7, „ 14, „ 10 — *For* $-\cdot 007066\, i_2 - \cdot 008148\, i_3$ *read* $+\cdot 006624\, i_2 + \cdot 008260\, i_3$.

Page 10, line 8, „ 13, „ 8, „ 14, „ 11 — *For* $-\cdot 008148\, i_2 + \cdot 001210\, i_3$ *read* $+\cdot 008260\, i_2 - \cdot 001238\, i_3$.

Page 14, line 3 from foot, *For* $-\cdot 02161$ *read* $-\cdot 00812$.

Page 15, line 7, *For* $-\cdot 4753$ *read* $+\cdot 4458$.

Page 16, lines 4, 5, *For* round its centre of mass *read* round the centre of mass of the earth and moon.

Page 19, equation (18), *For* $R=$ *read* $R=$real part of.

Page 69, line 2, *For* $3D - F + l + 2T - 3Q$ *read* $3D - F - l + 2T - 3Q$.

Page 75, line 11, *For* $+722$ *read* $+726$.
„ „ „ $-0''\cdot 007$ „ $-0''\cdot 003$.
„ „ „ -147 „ -148.

Page 77, column headed $\phi_0 = l + 3T - 10V$, *change the minus signs before* ϖ', ϖ'', h'' *to plus signs.*

Page 79, line 9, 1st column, *For* $+86$ *read* $+83$.
„ „ 10, „ „ $+5$ „ $+2$.
„ „ 8 from foot, *For* $\phi - 2F$ *read* $\phi + 2F$.

Page 80, line 1, *For* $3h + 24T$ *read* $3h + 22T$.
„ „ 4 from foot, 2nd column, *For* $+0''\cdot 029$ *read* $+0''\cdot 027$.
„ „ 3 „ „ *Delete* $\phi + l$ $+0''\cdot 002$.
„ „ 5 „ *For* $-0''\cdot 0152$ *read* $-0''\cdot 0086$.
„ last line, *For* ϕ $-0''\cdot 013$ *read* $\phi + 110^\circ$ $+0''\cdot 007$.

Printed in the United States
By Bookmasters